Maximal Cohen–Macaulay Modules and Tate Cohomology

Ragnar-Olaf Buchweitz, Luminy, 2004

Mathematical
Surveys
and
Monographs

Volume 262

Maximal Cohen–Macaulay Modules and Tate Cohomology

Ragnar-Olaf Buchweitz

With appendices by
Luchezar L. Avramov
Benjamin Briggs
Srikanth B. Iyengar
Janina C. Letz

AMERICAN
MATHEMATICAL
SOCIETY
Providence, Rhode Island

This book is based on an unpublished preprint, "Maximal Cohen-Macaulay modules and Tate-cohomology over Gorenstein rings," by Ragnar-Olaf Buchweitz, and is printed with permission by Ruth Buchweitz.

2020 *Mathematics Subject Classification.* Primary 13C14, 13Dxx, 13H10, 16Exx, 18G65.

For additional information and updates on this book, visit
www.ams.org/bookpages/surv-262

Library of Congress Cataloging-in-Publication Data

Names: Buchweitz, Ragnar-Olaf, 1952- author.
Title: Maximal Cohen-Macaulay modules and Tate cohomology / Ragnar-Olaf Buchweitz ; with appendices by Luchezar L. Avramov, Benjamin Briggs, Srikanth B. Iyengar, Janina C. Letz.
Description: Providence : American Mathematical Society, 2021. | Series: Mathematical surveys and monographs, 0076-5376 ; volume 262 | Includes bibliographical references and index.
Identifiers: LCCN 2021032204 | ISBN 9781470453404 (paperback) | 9781470467920(ebook)
Subjects: LCSH: Cohen-Macaulay modules. | Modules (Algebra) | Homology theory. | AMS: Commutative algebra – Theory of modules and ideals – Cohen-Macaulay modules [See also 13H10]. | Commutative algebra – Homological methods {For noncommutative rings, see 16Exx; for general categories, see 18Gxx}. | Commutative algebra – Local rings and semilocal rings – Special types (Cohen-Macaulay, Gorenstein, Buchsbaum, etc.) [See also 14M05]. | Associative rings and algebras {For the commutative case, see 13-XX} – Homological methods {For commutative rings, see 13Dxx; for general categories, see 18Gxx}.
Classification: LCC QA251.3 .B83 2021 | DDC 512/.44–dc23
LC record available at https://lccn.loc.gov/2021032204

10 9 8 7 6 5 4 3 2 1   26 25 24 23 22 21

Contents

Introduction

The main part of the book you are holding in your hand is a transcript of the preprint *Maximal Cohen–Macaulay modules and Tate cohomology over Gorenstein rings*, by Ragnar-Olaf Buchweitz. Buchweitz (1952–2017) was Professor of Mathematics at the University of Toronto, Canada. This preprint, which dates from 1986, was intended as a contribution to the conference proceedings:

Singularities, representation of algebras, and vector bundles (Lambrecht, 1985), Lecture Notes in Math., vol. 1273, Springer, Berlin, 1987, MR 915169.

While that did not come to pass, the preprint was circulated among a small circle of mathematicians working mainly in commutative algebra and in singularity theory. The central object of study, as the title suggests, is the category of maximal Cohen–Macaulay modules over Gorenstein rings. And one of the main results, according to Buchweitz himself, is that the stable category of maximal Cohen–Macaulay modules over a (not necessarily commutative) Gorenstein ring is equivalent to its stable derived category, namely, the Verdier quotient of the bounded derived category by the subcategory of perfect complexes. This assimilates and significantly extends earlier work of Eisenbud on hypersurface singularities, and of Tate on group cohomology of finite groups.

Over time, this preprint began to have an impact, albeit slowly at first, on research in commutative algebra. This changed in 2004 when Orlov rediscovered the stable derived category—calling it the "singularity category" in:

Triangulated category of singularities and D-branes in Landau–Ginzburg models, Trudy Steklov Math. Institute 246 (2004) 200–224. MR 2101296

Orlov's work lead to a renewed interest in Buchweitz's unpublished manuscript, and new generations of researchers have begun to discover the wealth of mathematics in it. A MathSciNet search for *Maximal Cohen–Macaulay modules and Tate cohomology over Gorenstein rings* in the References field on 21 June 2021 yielded 295 citations—the earliest from 2000 and 189 since 2015—to the typescript that has never been posted on `arXiv`. Buchweitz posted a scanned copy on the University of Toronto T-Space on 25 November 2008. It is available here:

`https://tspace.library.utoronto.ca/handle/1807/16682`

The first part of this book contains a newly typeset Preface, Chapters 0–10, and Bibliography of the original typescript; the page numbers in the original are recorded in the margins, in bold face. Only typographical and mathematical errors have been corrected; we found that most of these can be dealt with in the text, with changes indicated in footnotes when appropriate. The few cases of substantial mathematical issues are handled in Appendix A. Appendix B contains a self-contained proof of a version of Buchweitz's equivalence that covers also commutative rings of infinite Krull dimension. The last chapter, Appendix C, is a survey—limited by our interest and expertise—of some of the developments in the subjects related to, or spurred on by, Buchweitz's manuscript. The references pertinent to the appendices are collected in a separate bibliography and labeled by numbers to distinguish them from the original bibliographical entries. We have added an index and a glossary.

In preparing Ragnar's manuscript for publication we found that there is a host of ideas and suggestions, especially in later chapters, that remain to be explored. If this publication draws the community's attention to them, our purpose in undertaking the present task would be served.

It remains to thank Xiao-Wu Chen, Vincent Gélinas, Henning Krause, Graham Leuschke, and Jian Liu for their comments and suggestions. We are also grateful to the American Mathematical Society for publishing this manuscript, and especially to Sergei Gelfand for his interest and consistent encouragement.

We are thankful to Ruth Buchweitz for the permission to publish these notes and for providing Ragnar's picture that appears in the front matter. This is how we remember him.

Luchezar L. Avramov
Benjamin Briggs
Srikanth B. Iyengar
Janina C. Letz

Preface

The main theme of this article is: **1**[1]

Why should one consider maximal Cohen–Macaulay modules?

Although there has been a lot of work and success lately in the theory of such modules, of which this conference[2] witnessed, it has remained mysterious—at least to the present author—why these modules provide such a powerful tool in studying the algebra and geometry of singularities, for example.

We try to give one answer here, at least for the case of Gorenstein rings. Their role is special as over such rings "maximal Cohen–Macaulay" and "being a syzygy module of arbitrarily high order" are synonymous. It turns out that these modules, in a very precise sense, describe all *stable* homological features of such rings.

The motif was the observation that maximal Cohen–Macaulay modules—at least up to projective modules—carry a natural *triangulated structure* which implies that there is a naturally defined cohomology theory attached to these modules—the *Tate cohomology.*

To be more specific let us explain the essential points in the case of a local hypersurface ring R: It was observed by D. Eisenbud, [**Eis80**], that any finitely generated module over R admits a minimal free resolution which becomes eventually periodic of period two. Maximal Cohen–Macaulay modules over R—without free summands—are characterized as having a resolution periodic right from the start. Furthermore, the periodic part of the resolution comes from a "matrix factorization" of the defining equation and these matrix factorizations behave like "free complexes
modulo two", exhibiting the forementioned triangulated structure if one considers **2**
such matrix factorizations "up to homotopy".

Now one may proceed as follows: Trace back a minimal resolution of a module until the periodic tail is reached, turn around and extend by periodicity to a complex which is then unbounded, acyclic and consists only of free modules of finite rank.

If we started with a maximal Cohen–Macaulay module, except for possible free summands, no information is lost as the original module can be recovered from the image of a differential in this complex.

If we began with an arbitrary finitely generated R-module, the loss is the non-periodic part of the resolution, which is a finite free complex. In return we obtain a *complete resolution*—and a maximal Cohen–Macaulay module as its 0-th syzygy module—canonically attached to the (resolution of the) module we started with. More generally, one could have taken any complex of modules with bounded finitely

[1] This number refers to the page number in the manuscript published on the University of Toronto T-Space: `https://tspace.library.utoronto.ca/handle/1807/16682`

[2] See the first paragraph of the Introduction.

generated cohmology in the beginning to obtain still such a complete resolution, as well as a maximal Cohen–Macaulay module from "its" minimal resolution.

Checking the necessary details, which is rather straightforward, one has hence the following essentially equivalent data:

- Acyclic projective complexes up to homotopy, (that is, the "complete resolutions"),
- Maximal Cohen–Macaulay modules up to projective modules,
- Right bounded complexes with bounded, finitely generated cohomology, (the objects which are resolved), modulo bounded complexes of finitely generated projective modules, (the "heads" of the resolutions), all this is up to homotopies and quasi-isomorphisms.

The first and third of these categories carry natural triangulated structures compatible with the associations mentioned above. Hence they induce a natural structure on the middle one, which turns out to be the "triangulation" observed before. These three equivalent structures give rise to a cohomology theory, unbounded and stabilizing in the positive range the usual Ext-modules. (In case of hypersurfaces, this cohomology is furthermore by construction obviously periodic of period two.)

Carrying out this program, one observes immediately that the only fact needed on the ring—aside being noetherian—is that it has *finite injective dimension* as a module over itself, or, equivalently, that it admits a projective dualizing module. In
3 particular this means that *commutativity is in no way essential.* Hence we develop the theory for not necessarily commutative rings which are (strongly) *Gorenstein* in the sense that they are noetherian and of *finite injective dimension* as modules over themselves on both sides.

To define *maximal Cohen–Macaulay modules* (MCM for short) over such rings we use (one of) their defining properties in the commutative case: They will be those modules which are *acyclic with respect to the duality defined by* $\mathrm{RHom}_R(-, R)$.

Having extended the theory to include non-commutative "Gorenstein rings", quite different looking "classical" results become special cases of this same theory, for example:

- The theory of MCMs over a hypersurface ring as the starting point,
- The theory of integral group representations as initiated by J. Tate, (which has a common intersection with the above in the case of cyclic groups),
- The monadic description of certain (derived) categories of coherent sheaves of modules on varieties such as projective space—originally developed by Bernšteĭn–Gelfand–Gelfand and Beĭlinson, [**BGG78**], [**Gel84**]—or projective complete intersections defined by quadrics, obtained independently by Kapranov and the author. A more detailed account is contained in [**BEH87**, App.], this volume. (Here the intersection with the first theory consists of course of the projective quadrics and the general machine can be used—as was already implicitly done by R. G. Swan, [**Swa68**]—to determine the higher algebraic K-groups of smooth projective quadrics.)

The presentation of the general theory here is not unambiguous. Once the main idea is clear, one could essentially take any book on homological algebra, look at the chapter on Tate cohomology and generalise all its statements to the general case by a rather obvious dictionary.

As this is so, we choose a different path. We develop first the general theory in the framework of derived or triangulated categories to obtain the equivalence of categories sketched above and stated in precise form as a theorem in Chapter 4.

This we consider the main result.

From there, all the essential properties which are known for the Tate cohomology of finite groups follow more or less immediately in general and we try to make clear how it connects with the classical theory in Chapter 6.

But before that, in Chapter 5, we give some rather concrete applications to modules. It is shown that any module admits an *MCM-approximation*, that is, a 4
surjection from a maximal Cohen–Macaulay module onto the given one such that the kernel is of finite projective dimension. Such an MCM is uniquely determined up to free direct summands and yields a right adjoint of the inclusion of MCMs into the category of all finitely generated modules modulo projective modules. The same method—which is essentially the duality theorem for Gorenstein rings—yields then a method to construct for any complex a complete resolution by "symmetrisation". This will be used then in Chapter 7 to prove a duality theorem for Tate cohomology. It contains the classical theorem for integral group representations and our aim in Chapter 9 is to show that in fact it contains also Serre's duality theorem for projective space. For this we extend the theory of Bernšteĭn–Gelfand–Gelfand mentioned before in a straightforward manner to "linear super-spaces" and show that there the theory of maximal Cohen–Macaulay modules is essentially the same as the theory of coherent sheaves on the dual projective super-space. This is preceded in Chapter 8 by some easy examples, like minimal primitive quotients of enveloping algebras of semi-simple Lie-algebras or graded Lie-algebras—but for these rings the meaning of maximal Cohen–Macaulay modules—with respect to representation theory say—is not clear yet.

Concerning other applications, even for the most natural generalisation from hypersurfaces to complete intersections the general structure of the Tate cohomology groups is not known.[3] We offer only a result that bounds the support of these groups: It is clear from the construction that the support is in the non-regular locus, but more precisely, if there exists a Noether normalization, the corresponding Noether different will annihilate all Tate cohomology.

Hence, going back to hypersurfaces, we reverse the order of ideas and look at some peculiar features of Tate cohomology of cyclic groups and try to generalise them at least to hypersurfaces. The result is the "Herbrand difference" in Chapter 10 which generalises the classical notion of Herbrand quotients for cyclic groups. We develop its properties to convince the reader that it may serve as a suitable intersection form on ("co"-)primitive algebraic cycles in case of homogeneous polynomials whose underlying projective hypersurface is regular. Apart from rather simple examples like quadrics or cubic surfaces—where we regain the E_6-lattice algebraically—the use of this theory in connection with the description of algebraic cycles and the (variational) Hodge conjecture has still to be further pursued.

A final comment on the methods: Although the theory is non-commutative,
all the inspiration comes from the theory of commutative Gorenstein rings. This 5
has two disadvantages. First, some of the proofs can possibly be simplified by using more sophisticated non-commutative techniques. Secondly, some pre-requisites

[3] For subsequent development, see [**18**] and (C.5.9).

which are obvious in the commutative case—the duality for example—seem not to have been treated yet, so that we have to go through the general machinery once more. This makes in particular Chapter 7 a little bit cumbersome.

On the other hand, this enables us to avoid the use of more special structure—even in the case of group rings all results are established without using the co-algebra structure and cup-products.

Many people stimulated the work presented.

It will become apparent, I hope, how much this work was influenced by M. Auslander. Large parts here can be seen as special cases or applications of his "stable module theory". He essentially also ignited the recent interest in the theory of maximal Cohen–Macaulay modules and always insisted that it should be important. His remarks—sometimes sibylline, at least for the author—always brought about fruitful ideas.

I am also very grateful to L. Avramov who pointed out to me that what was termed "stable cohomology theory" then is indeed a natural extension of Tate's classical theory for finite groups. Without him, I would rather have missed this beautiful and clear connection.

G.-M. Greuel, H. Knörrer and F.-O. Schreyer helped through various discussions along the way and conversations with J. Steenbrink made me become aware of the geometric content of the theory.

I am particularly indebted to the organizers of this conference, G.-M. Greuel and G. Trautmann, for the opportunity to publish this work in the proceedings, although I was hindered from attending the actual conference. Also they showed incredible patience awaiting the final version of this manuscript.

CHAPTER 0

Notations and conventions

0.1 All rings considered will be associative and with unit, *left and right noetherian*, and all modules will be unitary—unless explicitly stated otherwise.

If S is any ring, $\mathrm{Mod}(S)$ ($\mathrm{mod}(S)$), denotes the category of all (finitely generated) *right* S-modules, and $\mathrm{P}(S)$ the full subcategory of all finitely generated projective S-modules.

If $S = \bigoplus S_i$ is a *graded* ring, $\mathrm{grMod}(S)$, $\mathrm{grmod}(S)$, $\mathrm{grP}(S)$, denote the corresponding categories of graded S-modules with degree-preserving S-linear maps as morphisms.

6

0.2 By $\mathrm{D}^*(S)$ for $*$ in $\{\ ,+,-,\mathrm{b}\}$ we denote the *derived categories* of $\mathrm{Mod}(S)$ whose objects are "all" complexes of S-modules with *finitely generated cohomology modules*[1], non-zero only in the range indicated by "$*$". (In Hartshorne's notation, [**Har66**], this would be $\mathrm{D}^*_{\mathrm{mod}(S)}(\mathrm{Mod}(S))$.)

Correspondingly, $\mathrm{K}^*(S)$ denotes the *homotopy categories* of those complexes of S-modules whose cohomology modules are again finitely generated. Each $\mathrm{D}^*(S)$ or $\mathrm{K}^*(S)$ is considered a *triangulated category* with respect to its natural triangulated structure.

In general, for terminology, notations and results on derived or triangulated categories, we use [**Ver77**] as a basic reference. In particular, let us mention explicitly the following conventions:

- *Complexes* will usually be indexed upstairs and will have differentials of degree $+1$: $X = (X, d_X) = (X^i, d_X^i\colon X^{i-1} \to X^i)_{i\in\mathbb{Z}}$.
- *Finitely generated S-modules* are tacitly identified with those complexes whose only non-zero term—if any—is in degree zero (in other words, the canonical embedding $\mathrm{mod}(S) \to \mathrm{D}^*(S)$ gets no name).
- In any triangulated category D the translation functor will be denoted T or $-[1]$, rather than T_{D}. In categories derived from complexes, it shifts one place "to the left", that is against the direction of the differential, and changes the sign of the differential while leaving morphisms "unchanged".
- Deviating from [**Ver77**] and following [**BBD82**, 1.1.1.], *distinguished triangles* will be displayed linearly:

$$X \xrightarrow{u} Y \xrightarrow{v} Z \xrightarrow{d} X[1]\,.$$

0.3 We will mostly deal with $\mathrm{D}^{\mathrm{b}}(S)$—which by definition now is the derived category of all complexes with finitely generated *total* cohomology. Without further notice it will be identified with its full triangulated subcategory of "projective

[1] The notation $\mathrm{D}(S)$ is usually reserved for the *full* derived category of all complexes of modules. The latter category does not play a significant role in this text.

resolutions", $\mathrm{K}^{-,\mathrm{b}}(\mathrm{P}(S))$, which is the homotopy category of all those bounded above complexes of finitely generated projective S-modules whose cohomology is bounded. An isomorphism $F \to X$ in $\mathrm{D}^{\mathrm{b}}(S)$ from an object F in $\mathrm{K}^{-,\mathrm{b}}(\mathrm{P}(S))$ is called a *projective resolution* of X.

CHAPTER 1

Perfect complexes and the stable derived category

7

1.1 A complex of S-modules is *perfect* if it is isomorphic in $\mathrm{D}(S)$ to a *bounded complex of finitely generated projective* S-modules.

Perfect complexes form an essential, (i.e. closed under isomorphisms), full and triangulated subcategory of $\mathrm{D^b}(S)$, denoted $\mathrm{D^b_{perf}}(S)$. The terminology "perfect" is borrowed from [**BGI71**, I, 2.1]—where it would be $\mathrm{P}(S)$-perfect—but it has nothing to do with the notion of a "perfect" module or ideal as defined by a grade condition—e.g. in [**Kap74**, p.126]. In the sense used here, *a module*—considered as a complex—*is perfect if and only if its projective dimension is finite.*

1.2 An intrinsic[1] characterization of perfect complexes is given by

LEMMA 1.2.1. *Let S be a ring, X an object in $\mathrm{D^b}(S)$. Then the following conditions are equivalent:*

(i) *X is perfect.*
(ii) *There is an integer $i(X)$ such that for any $i \geq i(X)$ and any finitely generated S-module M*

$$\mathrm{Ext}^i_S(X, M) := \mathrm{Hom}_{\mathrm{D^b}(S)}(X, \mathrm{T}^i M) = 0\,.$$

(iii) *For any exact functor $F\colon \mathrm{D^b}(S) \to \mathsf{D}$ into another triangulated category D for which $F(S) = 0$ one has $F(X) = 0$.*

PROOF. It follows from the definition that a complex X is perfect iff in (any of) its projective resolution(s) $F \to X$, all the *syzygy modules* $\mathrm{Cok}(d^i_F)$ "far enough back"—$i \ll 0$—are themselves projective modules. But as F has a priori bounded cohomology, this happens iff the augmented complex $(F^{\leqslant i} \to \mathrm{Cok}(d^i_F))$ is contractible for some i, iff condition (ii) is satisfied.

The equivalence of (i) and (ii) shows that $\mathrm{D^b_{perf}}(S)$ is a *thick triangulated subcategory* of $\mathrm{D^b}(S)$—in the sense of [**Ver77**, I.2.1.1.]. Condition (iii) on the other hand characterizes the *thick hull* of the single object S considered as subcategory of $\mathrm{D^b}(S)$. It is clear that this hull contains $\mathrm{D^b_{perf}}(S)$, which is already thick and contains S, whence one has in fact equality. This proves the equivalence of (i) and (iii). □

It follows from the characterization of $\mathrm{D^b_{perf}}(S)$ as a thick subcategory of $\mathrm{D^b}(S)$
that there exists a quotient, which is again triangulated and unique up to natural 8

[1]This is not truly intrinsic for it uses the t-structure on the derived category. For recent developments in this direction; see Neeman [**104**].

equivalence, satisfying the universal property that any functor F on $\mathrm{D}^{\mathrm{b}}(S)$ as in (iii) above factors through it.

To give the child a name, we make the

DEFINITION 1.2.2. The triangulated quotient category[2]

$$\underline{\mathrm{D}}^{\mathrm{b}}(S) := \mathrm{D}^{\mathrm{b}}(S)/\,\mathrm{D}^{\mathrm{b}}_{\mathrm{perf}}(S)$$

will be called the *stable derived category of* S[3]. The *projection functor* $\mathrm{D}^{\mathrm{b}}(S) \to \underline{\mathrm{D}}^{\mathrm{b}}(S)$ is the identity on objects, which hence still are complexes, and associates to a morphism f in $\mathrm{D}^{\mathrm{b}}(S)$—or to an actual morphism of complexes—its class $\underline{f}$ in $\underline{\mathrm{D}}^{\mathrm{b}}(S)$.

1.3 Remarks.

(a) "Stabilized" as by the lemma above the morphisms in $\underline{\mathrm{D}}^{\mathrm{b}}(S)$ satisfy[4]

$$\mathrm{Hom}_{\underline{\mathrm{D}}^{\mathrm{b}}(S)}(X, \mathrm{T}^i\, Y) = \mathrm{Hom}_{\mathrm{D}^{\mathrm{b}}(S)}(X, \mathrm{T}^i\, Y) = \mathrm{Ext}^i_S(X, Y)$$

for all $i \gg 0$—depending on X and Y.

(b) In practice, perfect complexes reflect those homological features of (complexes of) S-modules which are induced from "regular" rings in the following sense:

Assume S can be represented as a *left* module over some ring P of *finite right global dimension* (= "regular"). Then the image of the (left) derived functor

$$- \otimes^{\mathsf{L}}_P S\colon \mathrm{D}^{\mathrm{b}}(P) \longrightarrow \mathrm{D}^{\mathrm{b}}(S)$$

has as its thick hull $\mathrm{D}^{\mathrm{b}}_{\mathrm{perf}}(S)$. This follows as the assumptions just guarantee that the image of $\mathrm{D}^{\mathrm{b}}(P)$ under the (derived) tensor product consists only of perfect complexes and as it certainly contains (an object isomorphic to) S. Now one may conclude as in the proof of (i) $\Leftrightarrow$ (iii) above.

---

[2]Nowadays this would be called the Verdier quotient.

[3]Buchweitz called this the "stabilized derived category".

[4]This is not true without the Gorenstein hypothesis; see Remark A.1.1.

CHAPTER 2

The category of modules modulo projectives

2.1 Recall—from [**Hel60**, §3] or [**AB69**, 1.43] for example—the following

DEFINITION 2.1.1. The (projectively) *stable category*[1] of finitely generated S-modules, denoted $\underline{\mathrm{mod}}(S)$, is obtained by factoring out all the projective modules from $\mathrm{mod}(S)$. This means that $\underline{\mathrm{mod}}(S)$ has the *same objects* as $\mathrm{mod}(S)$ but that its morphisms are given by **9**

$$\begin{aligned}\underline{\mathrm{Hom}}_S(M,N) &:= \mathrm{Hom}_{\underline{\mathrm{mod}}(S)}(M,N)\\ &:= \frac{\mathrm{Hom}_S(M,N)}{\{f\colon M\to N \text{ factors over a projective } S\text{-module}\}}\end{aligned}$$

for all finitely generated S-modules M and N.

The category $\underline{\mathrm{mod}}(S)$ is still an additive category and S-modules M and N are isomorphic in $\underline{\mathrm{mod}}(S)$ if and only if they are *stably equivalent* (by projectives) in $\mathrm{mod}(S)$: $M\oplus P\cong N\oplus Q$ for some finitely generated projective S-modules P and Q—[**AB69**, 1.44].

2.2 On $\underline{\mathrm{mod}}(S)$ there is the *loop-space functor* Ω_S defined, introduced by Eckmann–Hilton and studied in general by A. Heller, [**Hel60**, 1]. It is obtained by choosing for any finitely generated S-module M a surjection from a finitely generated projective S-module $\mathsf{p}_M\colon \mathrm{P}_M\to M$ and setting

$$\Omega_S M := \mathrm{Ker}(\mathsf{p}_M)\,. \tag{2.2.1}$$

With these notations one has:

LEMMA 2.2.2. *The composition*

$$\mathrm{mod}(S)\longrightarrow \mathrm{D^b}(S)\longrightarrow \underline{\mathrm{D}}^{\mathrm{b}}(S)$$

factors uniquely over the canonical projection functor $\mathrm{mod}(S)\to\underline{\mathrm{mod}}(S)$ *and yields hence a naturally defined functor*

$$\iota_S\colon \underline{\mathrm{mod}}(S)\longrightarrow \underline{\mathrm{D}}^{\mathrm{b}}(S)\,.$$

It transforms the loop-space functor Ω_S into the inverse of the translation functor on $\underline{\mathrm{D}}^{\mathrm{b}}(S)$.

PROOF. Only the last assertion needs a proof. By definition of Ω_S there is a distinguished triangle

$$\iota_S(\Omega_S M)\longrightarrow \iota_S(\mathrm{P}_M)\longrightarrow \iota_S(M)\xrightarrow{\;d\;}\iota_S(\Omega_S M)[1]$$

in which $\iota_S(P_M)$ is obviously perfect. Hence the morphism d becomes an isomorphism in $\underline{\mathrm{D}}^{\mathrm{b}}(S)$. □

[1] Buchweitz called this the "stabilized category".

CHAPTER 3

Complete resolutions and the category of acyclic projective complexes

10

3.1 Adopting classical terminology—see for example [**CE99**, XII.3]—we make the following

Definition 3.1.1. Assume given a module M over some ring S. Then a *complete resolution of M* (over S) is an *acyclic complex*[1] (A, d_A) of *finitely generated projective* S-modules such that

$$\operatorname{Cok}(d_A^0 \colon A^{-1} \longrightarrow A^0) = M\,.$$

To abbreviate notations, the complex $A^- = (A^{\leqslant 0}, d_A|_{A^{\leqslant 0}})$ with its natural induced augmentation onto M is called the *associated projective resolution* of M, whereas the complex $A^+ = (A^{\geqslant 1}, d_A|_{A^{\geqslant 1}})[1]$ with its induced natural co-augmention from M into it is the *associated projective co-resolution* of M.

Remark that M is necessarily finitely generated if it admits a complete resolution and that a finitely generated module admits a complete resolution if and only if it admits a projective co-resolution. The complete resolution A itself is obtained as the *translated mapping cone* of $-d^0 \colon A^- \to A^+$, that is $A = \operatorname{cone}(-d^0)[1]$, so that $-d^0$ serves as the *connecting homomorphism* from the associated projective resolution to the associated projective co-resolution.

3.2 Instead of considering right away all those S-modules which admit a complete resolution—a seemingly hopeless task in general—we rather introduce the "category of complete resolutions" as an independent notion: $\underline{\mathrm{APC}}(S)$ denotes the *homotopy category of (unbounded) acyclic complexes* of *finitely generated projective S-modules.* (In Verdier's notation, [**Ver77**], this would be $\mathrm{K}^{\infty,\varnothing}(\mathrm{P}(S))$.) It is a full subcategory of $\mathrm{K}(S)$, closed under translation and forming mapping cones, hence *inherits a triangulated structure* form $\mathrm{K}(S)$.

3.3 On $\underline{\mathrm{APC}}(S)$ we define now the following functors: First, for *any* complex (X, d_X) of S-modules set

(3.3.1) $$\Omega_i(X) = \operatorname{Cok}(d_X^{-i} \colon X^{-i-1} \longrightarrow X^{-i}) \quad \text{for any integer } i$$

and call it the *i-th syzygy module* of X. As obviously 11

$$\Omega_0(X[1]) = \operatorname{Cok}(-d_X^1 \colon X^0 \longrightarrow X^1) = \Omega_{-1}(X)\,,$$

one obtains

(3.3.2) $$\Omega_i(X[j]) \cong \Omega_{i-j}(X)$$

[1] In this work "acyclic" means that every cohomology group is zero.

for all integers i, j and all complexes X. (Remark that this isomorphism of functors is not "canonical": For odd j it depends—at least—on the placement of $\pm\mathrm{id}$ in either even or odd degrees.)

If $f\colon X \to Y$ is a morphism of complexes of finitely generated projective S-modules which is *zero-homotopic*, then, for any i, the induced morphism of S-modules $\Omega_i(f)\colon \Omega_i(X) \to \Omega_i(Y)$ *factors over a finitely generated projective S-module*—namely even over *both* X^{-i+1} and Y^{-i}. This shows:

LEMMA 3.3.3. *Each Ω_i defines a functor from $\underline{\mathrm{APC}}(S)$ into $\underline{\mathrm{mod}}(S)$. It transforms the* inverse *of the translation functor on $\underline{\mathrm{APC}}(S)$ into the loop-space functor Ω_S on $\underline{\mathrm{mod}}(S)$.* □

(The reader may excuse the confusing use of Ωs here.)

3.4 Next we will set up functors from $\underline{\mathrm{APC}}(S)$ into $\underline{\mathrm{D}}^{\mathrm{b}}(S)$. For this recall that for any complex X its *naive filtration*, $(\sigma_{\leqslant k}X)_{k\in\mathbb{Z}}$, is given by
(3.4.1)

$$\begin{array}{ccccccccccccc}
\sigma_{\leqslant k+1}X := (\cdots & \longrightarrow & X^i & \longrightarrow & \cdots & \longrightarrow & X^k & \longrightarrow & X^{k+1} & \longrightarrow & 0 & \longrightarrow & \cdots) \\
 & & \big\downarrow{\scriptstyle \mathrm{id}_{X^i}} & & & & \big\downarrow{\scriptstyle \mathrm{id}_{X^k}} & & \big\downarrow & & \big\downarrow & & \\
\sigma_{\leqslant k}X \;:= (\cdots & \longrightarrow & X^i & \longrightarrow & \cdots & \longrightarrow & X^k & \longrightarrow & 0 & \longrightarrow & 0 & \longrightarrow & \cdots).
\end{array}$$

(Remark that the notation "$\sigma_{\leqslant k}$" for these "tronqués bêtes à droite" is in accordance with [**AGV72**, XVII,1.1.16] and [**BBD82**], but contrary to [**Har66**, I §7].)

One obviously has an equality of functors

$$\sigma_{\leqslant k} \circ \mathrm{T}^i = \mathrm{T}^i \circ \sigma_{\leqslant k+i} \quad \text{for all } k \text{ and } i\,. \tag{3.4.2}$$

12 Coming back to acyclic complexes of finitely generated projective S-modules, the following is easily established.

LEMMA 3.4.3. *Let A and B be objects of $\underline{\mathrm{APC}}(S)$. Then one has*

- (i) *In $\mathrm{D}(S)$ (the class of) the obvious morphism of complexes $\sigma_{\leqslant k}A \to (\Omega_{-k}A)[-k]$ becomes an isomorphism, or—equivalently—*
- (i') *$(\sigma_{\leqslant k}A)[k]$ is a projective resolution of $\Omega_{-k}A$.*
- (ii) *For any two integers $k \leq l$, the mapping cone over the natural morphism $\sigma_{\leqslant l}A \to \sigma_{\leqslant k}A$ is perfect.*
- (iii) *If $f\colon A \to B$ is a morphism of complexes which is homotopic to zero, all the induced morphisms $\sigma_{\leqslant k}f\colon \sigma_{\leqslant k}A \to \sigma_{\leqslant k}B$ have zero image in $\underline{\mathrm{D}}^{\mathrm{b}}(S)$. Hence these truncations induce well-defined functors*

$$\underline{\sigma_{\leqslant k}}\colon \underline{\mathrm{APC}}(S) \longrightarrow \underline{\mathrm{D}}^{\mathrm{b}}(S)\,.$$

PROOF. For (iii) just consider the commutative diagram of morphisms of complexes

$$\begin{array}{ccc}
\sigma_{\leqslant k}A & \longrightarrow & (\Omega_{-k}A)[-k] \\
\big\downarrow{\scriptstyle \sigma_{\leqslant k}f} & & \big\downarrow{\scriptstyle (\Omega_{-k}f)[-k]} \\
\sigma_{\leqslant k}B & \longrightarrow & (\Omega_{-k}B)[-k]
\end{array}$$

whose horizontal arrows become isomorphisms in $\mathrm{D}^{\mathrm{b}}(S)$—hence a fortiori in $\underline{\mathrm{D}}^{\mathrm{b}}(S)$—by (i). But, as observed above, f zero-homotopic implies that $\Omega_{-k}f$ factors over a projective module, and so the class of $(\Omega_{-k}f)[-k]$ is zero in $\underline{\mathrm{D}}^{\mathrm{b}}(S)$. □

These elementary facts show:

LEMMA 3.4.4. *The naive truncations $(\sigma_{\leqslant k})_{k\in\mathbb{Z}}$ define a directed system of functors*

$$\cdots \xrightarrow{\simeq} \underline{\sigma}_{\leqslant k} \xrightarrow{\simeq} \underline{\sigma}_{\leqslant k-1} \xrightarrow{\simeq} \cdots$$

from $\underline{\mathrm{APC}}(S)$ into $\underline{\mathrm{D}}^{\mathrm{b}}(S)$, whose transition morphisms are all isomorphisms. In particular, its inverse limit $\underline{\sigma}_{\leqslant} = \varprojlim_k \underline{\sigma}_{\leqslant k}$ exists and is an exact functor of triangulated categories

$$\underline{\sigma}_{\leqslant} \colon \underline{\mathrm{APC}}(S) \longrightarrow \underline{\mathrm{D}}^{\mathrm{b}}(S)\,.$$

CHAPTER 4

Maximal Cohen–Macaulay modules and Gorenstein rings

13

4.1 From now on, assume for the given ring S—which is still supposed to be noetherian on both sides—that:

S is of finite injective dimension both as a left and a right module over itself.

By a result of A. Zaks, [**Zak69**], if both the left and right injective dimensions of S are finite, they are the same and we will call this common value the *injective* or *virtual dimension* of S, and will abbreviate it as $\operatorname{vdim} S$. (Consequently we will occasionally say that S is a "ring of finite injective (virtual) dimension", if $\operatorname{vdim} S < \infty$.)

Such rings of finite virtual dimension could (and will) be called *(strongly) Gorenstein*[1] in view of the well-known "commutative"

THEOREM 4.1.1. *Assume that S is a noetherian commutative ring of finite Krull dimension. Then the following are equivalent:*

(i) *The injective dimension of S is finite.*
(ii) *For any prime $\mathfrak{p}$ of S, the localization $S_{\mathfrak{p}}$ is Gorenstein.*
(iii) *S admits a canonical module which is projective.*

Furthermore, under these conditions the injective dimension of S equals its Krull dimension. □

(i) $\Rightarrow$ (ii) is true for any noetherian commutative ring, but (ii) $\Rightarrow$ (i) needs the finite Krull dimension[2]. For this and (i) $\Leftrightarrow$ (iii) see, for example, [**67**, §1.4, Thm. 2] and [**FGR75**, 5.5, 5.6].

REMARK. We opt here for the term *strongly Gorenstein*, as in the commutative case it is indeed more restrictive than the usual definition which requires only (ii) above, but *not* the finite Krull dimension—see [**Har66**, p. 296] for example. On the other hand, M. Auslander introduced the notion of a (non-commutative) Gorenstein ring—see [**FGR75**, p. 47]—which, in general, does not imply "strongly Gorenstein". For example, as J.-E. Roos, [**Roo73**], pointed out, there are even rings of finite global dimension which are not Gorenstein in M. Auslander's sense. Nevertheless, using J.-E. Roos' results, many interesting (non-commutative) examples satisfy both definitions, namely those rings S which admit a filtration such that the 14
associated graded ring is commutative, Gorenstein and of finite Krull dimension.

[1]Such rings were earlier studied by Iwanaga in [**70**, **71**], and are sometimes called Iwanaga–Gorenstein rings.

[2]See Appendix B concerning the situation when $\dim S$ is not finite.

Let us also record the obvious fact that being strongly Gorenstein is left-right symmetric: *A ring S is strongly Gorenstein if and only if this holds for S^{op}*, the opposite ring of S.

4.2 We now come to the main object(s) of our study:

DEFINITION 4.2.1. Let S be a ring of finite injective dimension. Then a finitely generated S-module is *maximal Cohen–Macaulay* (MCM for short) if and only if

$$\operatorname{Ext}^i_S(M, S) = 0 \quad \text{for } i \neq 0\,.$$

The full subcategory of maximal Cohen–Macaulay modules in $\operatorname{mod}(S)$ is denoted $\operatorname{MCM}(S)$, and, accordingly, its image in $\underline{\operatorname{mod}}(S)$ by $\underline{\operatorname{MCM}}(S)$.

Again, the terminology is borrowed from commutative algebra, as over a local, commutative, Gorenstein ring it coincides with the usual notion. (Following M. Auslander—[**67**, 3.2.2], [**AB69**, Ch. 3]—one also could identify maximal Cohen–Macaulay modules as those of *G(orenstein)-dimension zero*. But, only dealing with (strongly) Gorenstein rings, we believe "MCM" to be more suggestive.)

(To keep the definition "coordinate-free" one may replace the module S in the above definition by any *faithfully projective* (right) S-module P: An object M in $\operatorname{mod}(S)$ is MCM if and only if

$$\operatorname{Ext}^i_S(M, P) = 0 \quad \text{for } i \neq 0\,,$$

see for example [**Bas68**, II.1.2].)

The analogy to the commutative case is supported as maximal Cohen–Macaulay modules in general share the following elementary properties:

LEMMA 4.2.2. *Let S be a ring which is strongly Gorenstein. Then*

(i) *Any finitely generated projective S-module is MCM, that is,* $\mathrm{P}(S)$ *is a subcategory of* $\operatorname{MCM}(S)$.

(ii) *If* $0 \to M_1 \xrightarrow{f} M_2 \to M_3 \to 0$ *is an exact sequence in* $\operatorname{mod}(S)$, *then*
- M_2, M_3 *in* $\operatorname{MCM}(S)$ *implies that* M_1 *is MCM,*
- M_1, M_3 *in* $\operatorname{MCM}(S)$ *implies that* M_2 *is MCM,*
15 - M_1, M_2 *in* $\operatorname{MCM}(S)$ *implies that* M_3 *is MCM iff* $\operatorname{Hom}_S(f, S)$ *is surjective.*

(iii) *A module M is MCM over S if*[3] *and only if* $M^* := \operatorname{Hom}_S(M, S)$ *is MCM as a right module over* S^{op}. *Furthermore, MCM's are reflexive:* $M \cong M^{**}$, *and a sequence of such is short exact in* $\operatorname{mod}(S)$, *if and only if the dual sequence in* $\operatorname{mod}(S^{\mathrm{op}})$ *is exact. In other words: The functor* $\operatorname{Hom}_S(-, S)$ *induces an exact equivalence between* $\operatorname{MCM}(S)$ *and* $\operatorname{MCM}(S^{\mathrm{op}})$.

(iv) *Any module in* $\operatorname{mod}(S)$ *admits a finite resolution by MCM's of length at most equal to* $\operatorname{vdim} S$. *(In such a resolution all but the last module can be chosen to be finitely generated projective.)*

PROOF. Everything, except perhaps the first assertion in (iii), is obvious from the definition and left to the reader.

That M is maximal Cohen–Macaulay over S iff M^* is so over S^{op} follows immediately by dualizing a projective resolution of M: By the very definition of MCM, this yields a projective co-resolution—(3.1)—of M^* in $\operatorname{mod}(S^{\mathrm{op}})$ and M^* is hence a syzygy module of arbitrarily high order. But as the (left) injective

[3]If M^* is MCM then M need not be MCM, unless it is reflexive. Only the forward direction in the proof is valid without this assumption.

dimension of S is finite, this shows further that M^* is necessarily $\mathrm{Hom}_S(-,S)$-acyclic, that is MCM. Furthermore, dualizing again, it follows that M is reflexive. □

4.3 Remarks.

(a) Again, the duality-statement (iii) above should rather be coordinate-free: Any *invertible S-bimodule ω_S* may serve as a *dualizing module*: The functor $\mathrm{Hom}_S(-,\omega_S)$ still defines an exact duality between $\mathrm{MCM}(S)$ and $\mathrm{MCM}(S^{\mathrm{op}})$, (c.f. [**Bas68**, II.5]).

An obvious advantage of such a coordinate-free description is that it behaves better functorially. As an example, it is left to the reader to convince themself—not needed in the sequel—that being "strongly Gorenstein" or "MCM" is invariant under Morita equivalence.

(b) In D. Quillen's terminology, [**Qui73**, I;§2], property (ii) above can be rephrased as follows:

$\mathrm{MCM}(S)$ is an *exact subcategory* of $\mathrm{mod}(S)$ in which *any* epimorphism is admissible, and in which the *admissible monomorphisms* are exactly those morphisms f which are monomorphic in $\mathrm{mod}(S)$ and *whose dual f^* is epimorphic* in $\mathrm{mod}(S^{\mathrm{op}})$.

In particular, there are defined *algebraic K-groups* for $\mathrm{MCM}(S)$ and property (iv) of the lemma shows—by Theorem 3, Corollary 3 of loc. cit.—that *the K-groups of* $\mathrm{mod}(S)$ *and* $\mathrm{MCM}(S)$ *are the same.*

16

4.4 Now we can state the main result[4]:

THEOREM 4.4.1. *Let S be a left and right noetherian ring, of finite injective dimension as a module over itself on either side. Then*

(1) *The syzygy functor Ω_0, defined in (3.3), induces an equivalence of categories—denoted by the same symbol—:*

$$\Omega_0\colon \underline{\mathrm{APC}}(S) \longrightarrow \underline{\mathrm{MCM}}(S)\,.$$

(2) *The restriction of ι_S, defined in (2.2), to $\underline{\mathrm{MCM}}(S)$ yields an equivalence of categories—again denoted by the same symbol—:*

$$\iota_S\colon \underline{\mathrm{MCM}}(S) \longrightarrow \underline{\mathrm{D}}^{\mathrm{b}}(S)\,.$$

(3) *The triangulated structures induced on $\underline{\mathrm{MCM}}(S)$ by either Ω_0 or ι_S are the "same" (in the sense that the identity on $\underline{\mathrm{MCM}}(S)$ becomes an exact isomorphism of triangulated categories), and—with respect to these structures—both functors are exact equivalences of triangulated categories, transforming the corresponding translation functor* T *into an inverse of the loop-space functor Ω_S, (2.2), restricted to $\underline{\mathrm{MCM}}(S)$.*

[4]Analogous statements hold for a larger class of (not necessarily strongly) Gorenstein rings; see Appendix B.

Before going into the proof, let us review the situation "graphically": There is a diagram of categories and functors, whose rows are "exact sequences" of categories

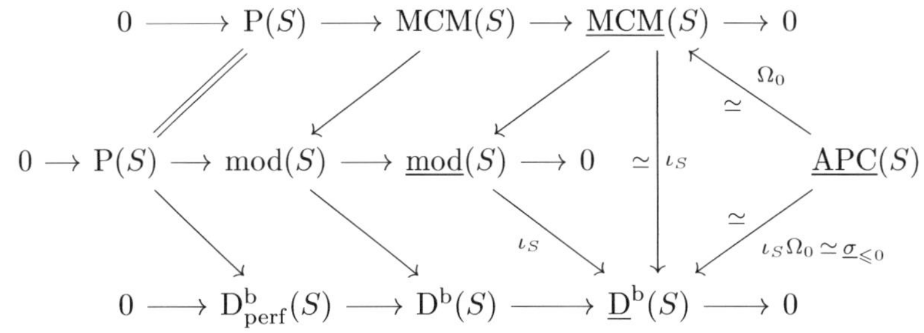

(the unlabeled morphisms are the canonical ones).

17

PROOF OF THE THEOREM. (i) *Ω_0 takes its values in $\underline{\mathrm{MCM}}(S)$.*

Assume given an acyclic complex A of finitely generated projective S-modules. Then

$$\mathrm{Ext}^i_S(\Omega_0 A, S) = \mathrm{Ext}^{i+j}_S(\Omega_{-j} A, S)$$

for all $i > 0$, $j \geq 0$. Now take $j > \operatorname{vdim} S - i$.

(ii) *Ω_0 is surjective on objects.*

Let M be a maximal Cohen–Macaulay module over S and $\mathbf{P}(M^*) \to M^*$ a projective resolution of M^* in $\mathrm{mod}(S^{\mathrm{op}})$. As in the proof of the lemma above,

$$0 \longrightarrow M^{**} \longrightarrow \mathrm{Hom}_{S^{\mathrm{op}}}(\mathbf{P}(M^*), S^{\mathrm{op}}) = \mathbf{P}(M^*)^*$$

will be a projective co-resolution of M^{**} in $\mathrm{mod}(S)$. Now $M \cong M^{**}$ as M is reflexive, and hence extending the projective co-resolution of M^{**} by a projective resolution of M yields a desired pre-image.

(iii) *Ω_0 is a full functor.*

Let $f\colon M \to N$ be an S-linear map of maximal Cohen–Macaulay modules over S. Extend it to a morphism $\mathbf{P}(f)\colon \mathbf{P}(M) \to \mathbf{P}(N)$ between chosen projective resolutions and analogously $f^*\colon N^* \to M^*$ to a morphism $\mathbf{P}(f^*)\colon \mathbf{P}(N^*) \to \mathbf{P}(M^*)$. Connecting $\mathbf{P}(M)$ and $\mathbf{P}(M^*)^*$ as well as $\mathbf{P}(N)$ and $\mathbf{P}(N^*)^*$ to complete resolutions of M and N respectively, $\mathbf{P}(f)$ and $\mathbf{P}(f^*)^*$ fit together to yield a morphism of these complete resolutions. By construction, this provides a pre-image of f, hence Ω_0 is full.

It remains to be seen that Ω_0 is faithful, to complete the proof of assertion (1). Instead of proving this—and assertion (2)—directly, we rather show:

$$\underline{\sigma}_{\leqslant}\colon \underline{\mathrm{APC}}(S) \longrightarrow \underline{\mathrm{D}}^{\mathrm{b}}(S) \textit{ is an equivalence of categories.}$$

This will readily imply the claim, since by (3.3), (3.4) there are natural isomorphisms of functors

$$\underline{\sigma}_{\leqslant} \xrightarrow{\simeq} \underline{\sigma}_{\leqslant 0} \xrightarrow{\simeq} \iota_S \circ \Omega_0\,,$$

whence $\underline{\sigma}_{\leqslant}$ being an equivalence gives that Ω_0 is faithful—and therefore also an equivalence by the above—so that finally also ι_S will be an equivalence of categories, establishing (2). So we prove:

18 (iv) *$\underline{\sigma}_{\leqslant}$ is essentially surjective.*

By (3.4), it will suffice to find for any complex X in $\mathrm{D}^{\mathrm{b}}(S)$ an integer k, an object A in $\underline{\mathrm{APC}}(S)$ and a morphism $X \to \sigma_{\leqslant k} A$ of complexes whose mapping cone is perfect. Also, replacing if necessary X by any of its resolutions, we may assume that X itself is already a (bounded above) complex of finitely generated projective S-modules. But then, all syzygy modules of X which sit "far enough back" have to be maximal Cohen–Macaulay. More precisely, using the same argument as (i) above,

$$\sigma_{\leqslant k}(X) = \operatorname{Cok}(d_X^k) = \operatorname{Im}(d_X^{k+1} \colon X^k \longrightarrow X^{k+1})$$

is certainly MCM for $k \leq \min(i : \mathrm{H}^i(X) \neq 0) - \operatorname{vdim} S$, and $\sigma_{\leqslant k}(X)[k]$ is a resolution of this module which can be extended to an acyclic complex of projectives by the argument already used in (ii) above.

(v) *$\underline{\sigma_{\leqslant}}$ is fully faithful.*

We use[5] Verdier's criterion—[**Ver77**, I.5.3]—: It suffices to prove that for a given perfect complex Y in $\mathrm{D^b}(S)$ and an object A in $\underline{\mathrm{APC}}(S)$ there exists an integer k such that all morphisms in $\mathrm{D^b}(S)$ from $\sigma_{\leqslant k}A$ into Y are zero. As $\sigma_{\leqslant k}A$ is a bounded above complex of projective modules, morphisms in $\mathrm{D^b}(S)$ from $\sigma_{\leqslant k}A$ into Y are in bijection with homotopy classes of (actual) morphisms of complexes and it is hence enough to show that any such morphism of complexes is indeed zero-homotopic.

Furthermore, it is enough to prove this assertion in case $Y = P[-i]$, where P is a finitely generated projective S-module and i is an integer, as these objects generate—up to isomorphism in $\mathrm{D^b}(S)$—any perfect complex by forming mapping cones.

Now, in this particular case, take any $k > i$ and let f be a morphism of complexes from $\sigma_{\leqslant k}A$ to $P[-i]$:

$$\begin{array}{ccccccccc}
\cdots & \longrightarrow & A^{i-1} & \xrightarrow{d^i} & A^i & \xrightarrow{d^{i+1}} & A^{i+1} & \longrightarrow & \cdots \\
 & & \downarrow & & \downarrow{\scriptstyle f^i} & & \downarrow & & \\
\cdots & \longrightarrow & 0 & \longrightarrow & P & \longrightarrow & 0 & \longrightarrow & \cdots .
\end{array}$$

The S-linear map f^i hence factors necessarily through $\operatorname{Cok}(d^i)$. Let $g \colon \operatorname{Cok}(d^i) \to P$ be the induced map. It remains to show that g can be further factored over the *inclusion* (by choice of k!) of $\operatorname{Cok}(d^i) = \Omega_{-i}A$ into A^{i+1}. But from the exact sequence

$$0 \longrightarrow \Omega_{-i}A \longrightarrow A^{i+1} \longrightarrow \Omega_{-i-1}A \longrightarrow 0$$

it follows that the obstruction for this lies in $\operatorname{Ext}^1_S(\Omega_{-i-1}A, P)$, which van- 19
ishes as P is finitely generated projective and $\Omega_{-i-1}A$ is still maximal Cohen–Macaulay by the argument in (i) above. (3) finally is just (2.2.2) and (3.3.3) reformulated. □

4.5 The proof shows how to find (quasi-)inverses of Ω_0 or ι_S:

Assume we have chosen projective resolutions $\mathbf{P}(M) \xrightarrow{p} M$ and $\mathbf{P}(M^*) \xrightarrow{q} M^*$ for any maximal Cohen–Macaulay module M over S. Then—by (ii) above and (3.1)—a *complete resolution of* M, denoted $\mathbf{CR}(M)$, is obtained by translating the mapping cone of the composition

$$\begin{aligned} d_M \colon \mathbf{P}(M) \xrightarrow{p} M \xrightarrow{\simeq} M^{**} \xrightarrow{q^*} \mathbf{P}(M^*)^* \\ \mathbf{CR}(M) := \operatorname{cone}(d_M)[-1], \end{aligned} \tag{4.5.1}$$

[5] This argument does not work since Verdier's criterion does not apply; see Remark A.4.1.

and one has by construction $\Omega_0(\mathbf{CR}(M)) = M$.

An *inverse of* ι_S involves:

- choosing for any complex X in $\underline{\mathrm{D}}^{\mathrm{b}}(S)$ a projective resolution $F \to X$,
- truncating this resolution at some
$$k \leq \min(i : \mathrm{H}^i(X) = \mathrm{H}^i(F) \neq 0) - \operatorname{vdim} S\,,$$
to obtain $\sigma_{\leqslant k}(F)$—for which then $\Omega_k(\sigma_{\leqslant k}F) = \Omega_k(F)$ is MCM,
- extending this truncated complex to an acyclic complex $\sigma_{\leqslant k}(F)^{\#}$ of finitely generated projective S-modules, and finally
- taking the 0-th syzygy module of this extension:

$$\iota_S^{-1}(X) = \Omega_0(\sigma_{\leqslant k}(F)^{\#})\,,$$
$$X \dashleftarrow F \xrightarrow{\sigma_{\leqslant k}} \sigma_{\leqslant k}(F) \overset{\sigma_{\leqslant k}}{\dashleftarrow} \sigma_{\leqslant k}(F)^{\#} \xrightarrow{\Omega_0} \iota_S^{-1}(X)\,. \tag{4.5.2}$$

Obviously, to find a representative of $\iota_S^{-1}(X)$ like this will be rather tedious in practice and we will give an often more useful device in the next section—at least in the case that X is (in $\mathrm{D}^{\mathrm{b}}(S)$ isomorphic to) just a single S-module—by exhibiting a *right adjoint* of the embedding $\underline{\mathrm{MCM}}(S) \to \underline{\mathrm{mod}}(S)$.

But first we use the description here to investigate the behavior of the just established equivalences with respect to *duality* and then to find the *distinguished triangles* for the induced exact structure on $\underline{\mathrm{MCM}}(S)$.

20

4.6 Let ω_S be a dualizing module for S as in Remark 4.3 (a). As ω_S is by definition finitely generated projective as an S-module,

- the functor $\mathrm{Hom}_S(-,\omega_S)$ on complexes transforms objects in $\underline{\mathrm{APC}}(S)$ into objects of $\underline{\mathrm{APC}}(S^{\mathrm{op}})$,
- the functor $\mathrm{Hom}_S(-,\omega_S)$ preserves the properties of being MCM or projective—apply (4.2.2) (iii) mutatis mutandis,
- the derived functor $\mathrm{RHom}_S(-,\omega_S)$ maps $\mathrm{D}^{\mathrm{b}}(S)$ into $\mathrm{D}^{\mathrm{b}}(S^{\mathrm{op}})$, as with S also ω_S is of finite injective dimension, and preserves the property of perfection.

Having observed all this, it follows that these functors induce exact dualities on each of the three categories involved in Theorem 4.4.1. More precisely:

PROPOSITION 4.6.1. *With notations as before, there is the following commutative diagram of triangulated categories and exact functors, whose rows are given by the equivalences of* (4.4.1) *for* S *and* S^{op} *respectively and whose vertical arrows are exact dualities induced by the functors above:*

$$\begin{array}{ccccc}
\underline{\mathrm{APC}}(S)^{\mathrm{op}} & \xrightarrow{\Omega_0{}^{\mathrm{op}}} & \underline{\mathrm{MCM}}(S)^{\mathrm{op}} & \xrightarrow{\iota_S{}^{\mathrm{op}}} & \underline{\mathrm{D}}^{\mathrm{b}}(S)^{\mathrm{op}} \\
\Big\downarrow{\scriptstyle \mathrm{Hom}_S(-,\omega_S)[-1]} & & \Big\downarrow{\scriptstyle \mathrm{Hom}_S(-,\omega_S)} & & \Big\downarrow{\scriptstyle \mathrm{RHom}_S(-,\omega_S)} \\
\underline{\mathrm{APC}}(S^{\mathrm{op}}) & \xrightarrow{\Omega_0} & \underline{\mathrm{MCM}}(S^{\mathrm{op}}) & \xrightarrow{\iota_{S^{\mathrm{op}}}} & \underline{\mathrm{D}}^{\mathrm{b}}(S^{\mathrm{op}})\,.
\end{array}$$

PROOF. The commutativity of the right-hand square expresses just the fact that maximal Cohen–Macaulay modules are by definition $\mathrm{Hom}_S(-,\omega_S)$-acyclic. For the square on the left, the foregoing construction of $\mathbf{CR}(-)$ as an inverse to the syzygy functor Ω_0 shows that for any maximal Cohen–Macaulay module M a complete resolution of its ω_S-dual $\mathrm{Hom}_S(M,\omega_S)$ can be obtained as

$$
\begin{aligned}
&\mathbf{CR}(\mathrm{Hom}_S(M, \omega_S)) && \\
&\quad \cong \mathrm{cone}(\mathrm{Hom}_S(d_M, \omega_S))[-1] && \text{by defintion of } \mathbf{CR}(-) \text{ and} \\
& && \text{the fact that on } \mathrm{MCM}(S) \\
& && \mathrm{Hom}_{S^{\mathrm{op}}}(\mathrm{Hom}_S(-, \omega_S), \omega_S) \cong \mathrm{id}; \\
&\quad \cong \mathrm{cone}(\mathrm{Hom}_S(d_M, \omega_S)[-1]) && \text{by definition of mapping cones;} \\
&\quad \cong \mathrm{Hom}_S(\mathrm{cone}(d_M), \omega_S) && \text{by a (non-canonical)} \\
& && \text{isomorphism of complexes;} \\
&\quad = \mathrm{Hom}_S(\mathrm{cone}(d_M)[-1], \omega_S)[-1] && \text{obviously;} \\
&\quad = \mathrm{Hom}_S(\mathbf{CR}(M), \omega_S)[-1] && \text{by definition of } \mathbf{CR}(M)\,.
\end{aligned}
$$ **21**

Applying Ω_0 then yields the result.

(In less formal terms, the shift by $[-1]$ on the left is caused by the fact that connecting a projective resolution of M with the dual of one for M^* "naively", creates a complex indexed by the integers with zero "doubled". Re-indexing "correctly"—as in (3.1)—introduces the shift if one dualizes.) □

4.7 Next we want to describe the triangulated structure on $\underline{\mathrm{MCM}}(S)$ directly. In general, such a structure on an additive category is determined by its *distinguished triangles* and we will use the theorem to determine those in the induced structure on $\underline{\mathrm{MCM}}(S)$.

For this, let $f\colon M \to N$ be any S-linear map of maximal Cohen–Macaulay S-modules. Choose an embedding $i\colon M \to Q$ of M into a finitely generated projective S-module such that its cokernel is still MCM. (This means just that Q serves as the first term in a projective co-resolution of M.) Then define a *mapping cone* $\mathrm{cone}(f)$ of f as the push-out—or amalgamated sum—of f and i, so that there is a commutative diagram of short exact sequences of S-modules

(4.7.1)
$$
\begin{array}{ccccccccc}
0 & \longrightarrow & M & \xrightarrow{\ i\ } & Q & \xrightarrow{\ p\ } & \mathrm{Cok}(i) & \longrightarrow & 0 \\
 & & {\scriptstyle f}\downarrow & & \downarrow & & \| & & \\
0 & \longrightarrow & N & \xrightarrow{\ i'\ } & \mathrm{cone}(f) & \xrightarrow{\ p'\ } & \mathrm{Cok}(i) & \longrightarrow & 0\,.
\end{array}
$$

Remark that $\mathrm{Cok}(i)$ represents $\mathrm{T}(M)$, the translate of M, as by (2.2.1) $\Omega_S(\mathrm{Cok}\, i)$ is represented by M and by (4.4.1) (3) the translation functor T on $\underline{\mathrm{MCM}}(S)$ is an inverse of Ω_S. Now call

(4.7.2)
$$M \xrightarrow{\ f\ } N \xrightarrow{\ i'\ } \mathrm{cone}(f) \xrightarrow{\ -p'\ } \mathrm{T}(M) = \mathrm{Cok}(i)$$

a *typical triangle*. Then the distinguished triangles in $\underline{\mathrm{MCM}}(S)$ are given by all those sequences of morphisms in $\underline{\mathrm{MCM}}(S)$ which are isomorphic to the image of a typical triangle under the projection functor $\mathrm{MCM}(S) \to \underline{\mathrm{MCM}}(S)$.

That this yields in fact exactly all distinguished triangles in $\underline{\mathrm{MCM}}(S)$ is most **22**
easily seen by applying Ω_0 to the distinguished triangles in $\underline{\mathrm{APC}}(S)$.

4.8 Remark. Starting with (4.7.1)–4.7.2 to *define* distinguished triangles on $\underline{\mathrm{MCM}}(S)$, it is prossible—but lengthy—to verify directly the axioms—(TR1) up to (TR4) in [**Ver77**, I.1.1]—defining a triangulated category.[6]

---

[6]This approach is discussed in [**65**] and [**39**].

The key fact, from this point of view, is that MCM(S) is an *exact subcategory* of mod(S), (see Remark 4.3 (b)), which is *Frobenius*:

This means—following A. Heller [**Hel60**, §3]—that with respect to the given exact structure there exists an object—namely S in MCM(S)—serving at the same time both as a (small, admissible) *projective generator*—which then yields by the usual procedure (admissible) projective resolutions—*and* as a (small, admissible) *injective cogenerator*—which hence allows the construction of (admissible) projective co-resolutions.

From these assumptions—*exact* and *Frobenius*—it follows already that $\underline{\mathrm{MCM}}(S)$, the stable category obtained by dividing out—in the sense of (2.1)—the projective-injective objects, is *triangulated* in a natural way, the distinguished triangles being defined as above.

For example, the most laborious axiom of a triangulated structure, namely (TR2) of (loc. cit.), which says that distinguished triangles can be rotated, was already anticipated by A. Heller, [**Hel60**, Thm.5.3], where it was (just?) a "remarkable property"—triangulated structures were not defined yet! (That "Frobenius" was the right categorical notion in our context was pointed out to me by D. Happel, who also described other examples arising from the representation theory of artinian algebras—see [**Hap85**].)

Continuing in this abstract setting, one may wonder for which morphisms of triangulated categories—as the embedding $\mathrm{D}^{\mathrm{b}}_{\mathrm{perf}}(S) \to \mathrm{D}^{\mathrm{b}}(S)$ or the derived tensor product $- \otimes^{\mathsf{L}}_{P} S\colon \mathrm{D}^{\mathrm{b}}(P) \to \mathrm{D}^{\mathrm{b}}(S)$ of the Remark 1.3 (b)—an "excision theorem" like (4.4.1) (2) can be established.

As a first step in this direction, one has of course to know how to recognize the "excised" subcategory, which is MCM(S) as a full subcategory of $\mathrm{D}^{\mathrm{b}}(S)$ in our case. Here, this can be done as follows—see [**BBD82**] for the no(ta)tions used:

23 Both $\mathrm{D}^{\mathrm{b}}(S)$ and $\mathrm{D}^{\mathrm{b}}(S^{\mathrm{op}})$, the derived category of left S-modules, carry natural *t-structures* whose *hearts* are precisely (equivalent to) the abelian category mod(S), respectively mod(S^{op}), which were "derived".

Now a "*dualizing complex*"—which in this case is just S or, more generally, any dualizing module ω_S, (4.3) (a)—establishes an exact duality

$$\mathrm{RHom}_S(-,\omega_S)\colon \mathrm{D}^{\mathrm{b}}(S)^{\mathrm{op}} \longrightarrow \mathrm{D}^{\mathrm{b}}(S^{\mathrm{op}})\,,$$

as already used in (4.6). Pulling back the natural t-structure from $\mathrm{D}^{\mathrm{b}}(S^{\mathrm{op}})^{\mathrm{op}}$ onto $\mathrm{D}^{\mathrm{b}}(S)$ by this duality establishes hence a second such structure on $\mathrm{D}^{\mathrm{b}}(S)$, whose "heart"—equivalent to mod(S^{op}) by construction—consists of all those complexes X of (right) S-modules for which the ω_S-dual $\mathrm{RHom}_S(X,\omega_S)$ has its only non-vanishing cohomology in degree zero.

The intersection in $\mathrm{D}^{\mathrm{b}}(S)$ of the hearts of these two t-structures is by the very definition equivalent to its full subcategory of all maximal Cohen–Macaulay modules over S, that is MCM(S)—endowed already with its exact structure.

(4.8.1) It seems reasonable to expect that duality theory is at the heart of the matter—and at least for "rings with dualizing modules", that is the (not necessarily commutative) generalisation of Cohen–Macaulay rings, one may indeed extend (4.4.1) (2). The price to pay is some more categorical machinery as one needs a more complicated stabilized version of $\underline{\mathrm{MCM}}(S)$—one has then also formally to invert the still existing "loop-space functor", (2.2), on the category of MCM's modulo projectives.

4.9 We close this section with a remark on K-groups.

Following D. Quillen [**Qui73**], we denote $\mathsf{G}_i(S) := \mathsf{K}_i(\mathrm{mod}(S))$[7] the algebraic K-groups of all finitely generated (right) S-modules and $\mathsf{K}_i(S) := \mathsf{K}_i(\mathrm{P}(S))$ the groups obtained from all finitely generated projective S-modules. On the other hand, as for any triangulated category—[**Ill77**, VIII]—there is defined the Grothendieck group of $\underline{\mathrm{MCM}}(S)$, which will be denoted $\underline{\mathsf{K}}_0(S)$. Then one has:

- an exact sequence $\mathsf{K}_0(S) \to \mathsf{G}_0(S) \to \underline{\mathsf{K}}_0(S) \to 0$ by [**Ill77**, VIII.3.1] and (4.4.1) (2), as well as
- group homomorphisms $\mathsf{K}_i(S) \to \mathsf{G}_i(S)$ for any i by [**Qui73**, §7].

Hence the following question seems rather natural: **24**

Given a ring S which is strongly Gorenstein, does there exist a “stabilized” higher algebraic K-theory for S, in the sense that there are naturally defined groups $\underline{\mathsf{K}}_i(S) = \mathsf{K}_i(\underline{\mathrm{MCM}}(S))$ for $i > 0$ which fit the short exact sequence and group homomorphisms just mentioned into a long exact sequence

$$\cdots \longrightarrow \underline{\mathsf{K}}_{i+1}(S) \longrightarrow \mathsf{K}_i(S) \longrightarrow \mathsf{G}_i(S) \longrightarrow \underline{\mathsf{K}}_i(S) \longrightarrow \cdots ?$$

As will be seen later in Chapter 10 in some special cases such groups can be defined, but starting from rather different interpretations of $\underline{\mathrm{MCM}}(S)$ and its equivalent companions.[8]

---

[7] Buchweitz wrote K' instead of G. We adopt the current standard.

[8] It is possible to define a stable version of the K-groups in general; see for example [**110**].

CHAPTER 5

Maximal Cohen–Macaulay approximations

5.0 From now on, it will be assumed—if not explicitly stated otherwise—that modules are finitely generated and that complexes of modules have finitely generated, bounded cohomology.

The aim of this section is to investigate how the subcategory of maximal Cohen–Macaulay modules is embedded into the category of all finitely generated modules.

As a more practical aspect of the general theory developed so far, we will give three (essentially equivalent) presentations of an arbitrary module in terms of a maximal Cohen–Macaulay module and a module of finite projective dimension. This extends (and follows) results of M. Auslander, [**67**], and M. Auslander–M. Bridger, [**AB69**].

5.1 We start with the following lemma which shows that over a ring S, which is noetherian and of finite injective dimension, maximal Cohen–Macaulay modules and modules of finite projective (or injective) dimension are "orthogonal" with respect to Ext_S.

LEMMA 5.1.1. *Let S be a ring which is strongly Gorenstein. Then*

(i) *An S-module M is MCM iff $\mathrm{Ext}^i_S(M,U) = 0$ for $i \neq 0$ and all S-modules U of finite projective dimension.*

(ii) *An S-module U is of finite projective dimension iff $\mathrm{Ext}^i_S(M,U) = 0$ for $i \neq 0$ and all MCM S-modules M.*

(iii) *An S-module U is of finite projective dimension iff it is of finite injective dimension.*

(iv) *An S-module M is MCM and of finite projective dimension iff it is projective.* **25**

(v) *Any S-linear map from an MCM to a module of finite projective dimension factors through a projective module.*

PROOF. In (i), (resp. (iv)), the "if" part follows from the definition of MCM (and (4.2.2) (i)), the "only if" part by induction on the projective dimension of U (resp. M)[1].

In (ii) again, the "only if" part is obtained by induction on the projective dimension of U, whereas the "if" part follows from (iv) and (4.2.2) (iv): Take M to be the "last" module in a finite resolution of U, in which all modules are MCM and all except perhaps M are projective. Then M is necessarily both MCM and of finite projective dimension, hence itself projective.

Using (4.2.2) (iv) once more, it is clear that (ii) and (iii) are equivalent.

Finally, assertion (v) follows from (i) as well as (ii). □

The next result shows that in fact maximal Cohen–Macaulay modules together with those of finite projective (injective) dimension "span" the category $\mathrm{mod}(S)$.

[1] The "only if" direction requires some explanation; see Remark A.5.1.

THEOREM 5.1.2 (The syzygy-theorem for Gorenstein rings). *Let S be a ring which is strongly Gorenstein.*

(1) *Any finitely generated S-module N admits a presentation*

$$0 \longrightarrow U \longrightarrow M \xrightarrow{p} N \longrightarrow 0, \tag{5.1.3}$$

where U is of finite projective dimension and M is maximal Cohen–Macaulay.

(2) *Such a presentation is unique up to (projectively) stable equivalence: If $N = M_1/U_1$ is a second presentation of the same kind, there exist finitely generated projective S-modules P and Q such that $M \oplus P \cong M_1 \oplus Q$ and $U \oplus P \cong U_1 \oplus Q$.*

(3) *The surjection $p\colon M \to N$ is universal with respect to the following property: Whenever $f\colon M' \to N$ is a morphism from some MCM S-module M' to N, it factors over p and this factorization is unique in $\underline{\mathrm{mod}}(S)$.*

An essentially equivalent form of this theorem is:

26

THEOREM 5.1.4. *Let S be a ring which is strongly Gorenstein.*

(1) *Any finitely generated S-module N occurs in an exact sequence*

$$0 \longrightarrow N \xrightarrow{i} V \longrightarrow L \longrightarrow 0, \tag{5.1.5}$$

where V is of finite projective dimension, and L is maximal Cohen–Macaulay.

(2) *Such a presentation is unique up to (projectively) stable equivalence in the sense above.*

(3) *The injection $i\colon N \to V$ is universal with respect to the following property: Whenever $j\colon N \to V'$ is a morphism from N to some V' of finite projective dimension, it factors over i and this factorization is unique in $\underline{\mathrm{mod}}(S)$.*

(4) *Comparing with the representation in* (5.1.3), *L is isomorphic to $\mathrm{T}(M)$, the translate of M, in $\underline{\mathrm{MCM}}(S)$.*

PROOF OF (5.1.2) AND (5.1.4). The asserted uniqueness in 2 as well as the claimed universal mapping property in 3—which statements are easily seen to be equivalent anyway—follow immediately from the lemma above. One also may deduce these properties—together with (5.1.4) 4—from Theorem 4.4.1: Converting as usual the exact sequence in (5.1.3) and (5.1.5) into distinguished triangles in $\mathrm{D^b}(S)$ and then projecting them into the stable derived category $\underline{\mathrm{D}}^{\mathrm{b}}(S)$, U resp. V become zero-objects, whence one obtains isomorphisms in $\underline{\mathrm{D}}^{\mathrm{b}}(S)$:

$$U \cong 0 \longrightarrow M \xrightarrow[\simeq]{p} N \longrightarrow 0 \cong U[1] \qquad \text{and}$$

$$V \cong 0 \longrightarrow L \xrightarrow[\simeq]{} N[1] \longrightarrow 0 \cong V[1].$$

Furthermore, as $\iota_S\colon \underline{\mathrm{MCM}}(S) \to \underline{\mathrm{D}}^{\mathrm{b}}(S)$ is an exact equivalence, this shows that M resp. L are unique up to isomorphisms in $\underline{\mathrm{MCM}}(S)$: M represents $\iota_S^{-1}(N)$ and L represents $\iota_S^{-1}(N[1]) = (\iota_S^{-1}(N))[1] = \mathrm{T}(M)$.

Hence it remains to verify the existence of the claimed representations. For this let $F \to N$ be a projective resolution of N.

As S is of finite injective dimension, the complex $F^* := \mathrm{Hom}_S(F, S)$ has still bounded cohomology. Now let $\varphi\colon G \to F^*$ be a projective resolution (over S^{op} of course) of F^*—remark that although F^* is a complex of projective S-modules, it is in general not bounded above!

Dualising again, one obtains a quasi-isomorphism[2] from $F \cong F^{**}$ to a bounded below complex G^* of finitely generated projective S-modules:
(5.1.6)

$$\begin{array}{ccccccccccccc}
\cdots F^{-k-1} & \longrightarrow & F^{-k} & \longrightarrow \cdots \longrightarrow & F^{-1} & \longrightarrow & F^{0} & \longrightarrow & 0 & & & = F \cong F^{**} \\
\downarrow & & \downarrow & & \downarrow & & \downarrow & & \downarrow & & & {\scriptstyle =\varphi^*}\downarrow \\
0 & \longrightarrow & (G^k)^* & \longrightarrow \cdots \longrightarrow & (G^1)^* & \xrightarrow{\delta^0} & (G^0)^* & \xrightarrow{\delta^1} & (G^{-1})^* & \xrightarrow{\delta^2} & \cdots & = G^*
\end{array}$$ 27

where we have set $\delta^j := \mathrm{Hom}_S(d_G^{-j+1}, S)$ for all j.

Remark that the largest k for which $G^k \neq 0$ can be chosen to be

$$k := \max(i : \mathrm{Ext}^i_S(N, S) \neq 0)\,. \tag{5.1.7}$$

As φ^* is a quasi-isomorphism, the only cohomology of G^* is the given module N in degree zero. Now set:

$$\begin{array}{lllll}
M := \mathrm{Ker}(\delta^1) & \text{and} & U := \mathrm{Im}(\delta^0)\,, & \text{so that} & N = M/U\,, \\
V := \mathrm{Cok}(\delta^0) & \text{and} & L := \mathrm{Im}(\delta^1)\,, & \text{so that} & N = \mathrm{Ker}(V \xrightarrow{\delta'} L)\,,
\end{array}$$

where δ' is the surjection induced from δ^1.

It follows from the construction that U and V are of finite projective dimension, as they are resolved respectively by the "negative (non-positive) tail" of G^*, whereas M and L are necessarily MCM as they allow respectively the "non-negative (positive) head" of G^* as projective co-resolutions. Hence we have found the desired representations of N.

Remark also that this construction yields another, direct, proof of (5.1.4) 4: M and L occur in the short exact sequence

$$0 \longrightarrow M = \mathrm{Ker}(\delta^1) \longrightarrow (G^0)^* \longrightarrow \mathrm{Im}(\delta^1) = L \longrightarrow 0 \tag{I}$$

which shows, using (4.7), that $L = M[1]$ in $\underline{\mathrm{MCM}}(S)$. □

5.2 Remarks.

(a) In a presentation as in (5.1.3), U cannot be an arbitrary module of finite projective dimension as it embeds into the MCM M. More precisely, the definition of U shows that it occurs in the short exact sequence

$$0 \longrightarrow U = \mathrm{Im}(\delta^0) \longrightarrow (G^0)^* \longrightarrow \mathrm{Cok}(\delta^0) = V \longrightarrow 0\,, \tag{II}$$

whence it is already the first syzygy module of the module V of finite projective dimension. Applying $\mathrm{Ext}_S(-, S)$ to (5.1.3), (5.1.5), one obtains furthermore

$$\mathrm{Ext}^i_S(U, S) = \mathrm{Ext}^{i+1}_S(N, S) = \mathrm{Ext}^{i+1}_S(V, S) \quad \text{for} \quad i > 0\,.$$

(b) As communicated to us by G. Evans, he and P. Griffith independently obtained (5.1.2) in the commutative case. 28

[2] This requires some explanation; see A.5.2.

5.3 The four exact sequences (5.1.3), (5.1.5), (I) and (II) fit into the following commutative diagram:

$$
(5.3.1)\qquad
\begin{array}{ccccccccc}
 & & 0 & & 0 & & & & \\
 & & \uparrow & & \uparrow & & & & \\
0 & \longrightarrow & N & \xrightarrow{\ i\ } & V & \xrightarrow{\ \delta'\ } & L & \longrightarrow & 0 \qquad (5.1.5)\\
 & & {\scriptstyle p}\uparrow & & {\scriptstyle p'}\uparrow & (*) & \| & & \\
0 & \longrightarrow & M & \xrightarrow{\ i'\ } & (G^0)^* & \longrightarrow & L & \longrightarrow & 0 \qquad (\mathrm{I})\\
 & & \uparrow & (*) & \uparrow & & & & \\
 & & U & = & U & & & & \\
 & & \uparrow & & \uparrow & & & & \\
 & & 0 & & 0 & & & & \\
 & & (5.1.3) & & (\mathrm{II}) & & & &
\end{array}
$$

where i' is the natural inclusion induced by δ^1 and p' the natural surjection induced by δ^0.

To interpret this diagram in terms of triangulated categories, consider its image in $\mathrm{D}^{\mathrm{b}}(S)$. It is then just the "display" of the octahedron constructed—by choice—either over the composable monomorphisms

$$U = \mathrm{Im}(\delta^0) \longrightarrow M = \mathrm{Ker}(\delta^1) \xrightarrow{\ i'\ } (G^0)^*\,,$$

or the composable epimorphisms

$$(G^0)^* \xrightarrow{\ p'\ } V = \mathrm{Cok}(\delta^0) \xrightarrow{\ \delta'\ } L = \mathrm{Im}(\delta^1)\,.$$

The distinguished triangles of the octahedron are given by the four exact sequences above, its two "visible" commutative faces are marked by $(*)$, its "top" is L, its "bottom" is U and the "equator" is represented by the square (i', p', i, p)—see [**BBD82**, 1.1.7.–8.]. Accordingly, we will call (5.3.1) "the" *canonical representation* or *octahedron* associated to a module N over a strongly Gorenstein ring S. It is
29 unique up to (projectively) stable equivalence in the sense that to any one of the "triangles" $(*)$ one may add one and the same projective module to two of the three vertices.

Remark that the "equator" is a bi-cartesian square which yields hence a Mayer–Vietoris sequence

$$(5.3.2)\qquad 0 \longrightarrow M \xrightarrow{\ p\oplus i'\ } N \oplus (G^0)^* \xrightarrow{\ i-p'\ } V \longrightarrow 0\,,$$

exhibiting the given module N—up to adding a projective module—as an extension of a module of finite projective dimension by an MCM. Again this representation of N is unique in $\underline{\mathrm{mod}}(S)$, proving the so-called "Approximation Theorem" of M. Auslander [**67**, Ch. 3, Prop. 8], [**AB69**, 4.27, 2.41]:

COROLLARY 5.3.3. *With the notations of* (5.1.2)*, there exist a finitely generated projective S-module Q and an exact sequence*

$$0 \longrightarrow M \longrightarrow N \oplus Q \longrightarrow V \longrightarrow 0\,,$$

in which M is a maximal Cohen–Macaulay module and V is of finite projective dimension, so that in $\underline{\mathrm{mod}}(S)$

- *the morphism $M \to N \oplus Q \cong N$ is universal with respect to morphisms from MCMs to N, and*
- *the morphism $N \cong N \oplus Q \to V$ is universal with respect to morphisms from N into modules of finite projective dimension.*

PROOF. Take $Q = (G^0)^*$ and rewrite (5.1.2), (5.1.4). □

5.4 Following a suggestion of M. Auslander, we give the objects just constructed the following names:

DEFINITION 5.4.1. Given an S-module N and its canonical representation (5.3.1), the occurring maximal Cohen–Macaulay module M (with the projection p onto N) will be called a *MCM-approximation* of N. Accordingly, the module V (with the embedding ι of N into it) is a *hull of finite projective dimension* of N.

Choosing MCM-approximations (hulls of finite projective dimension) for any S-module yields a functor, denoted $\mathbf{M}$ (resp. $\mathbf{H}$) from $\underline{\mathrm{mod}}(S)$ to $\underline{\mathrm{MCM}}(S)$ (resp. $\underline{\mathrm{fpd}}(S)$, the full subcategory of $\underline{\mathrm{mod}}(S)$ spanned by all modules of finite projective dimension).

In terms of these functors, the universal mapping properties (5.1.2) 3 and **30**
(5.1.4) 3 express of course just that $\mathbf{M}$ yields a *right adjoint* of the embedding $\underline{\mathrm{MCM}}(S) \to \underline{\mathrm{mod}}(S)$, whereas $\mathbf{H}$ defines a *left adjoint* of the embedding $\underline{\mathrm{fpd}}(S) \to \underline{\mathrm{mod}}(S)$.

Summarising the above results "categorically", one may say that $\underline{\mathrm{mod}}(S)$ is *almost obtained by gluing* $\underline{\mathrm{MCM}}(S)$[3] and $\underline{\mathrm{fpd}}(S)$ along the pairs of adjoint functors:

$$\underline{\mathrm{MCM}}(S) \underset{\mathbf{M}}{\overset{\text{inclusion}}{\rightleftarrows}} \underline{\mathrm{mod}}(S) \underset{\text{inclusion}}{\overset{\mathbf{H}}{\rightleftarrows}} \underline{\mathrm{fpd}}(S) \tag{5.4.2}$$

if compared to the notion of "gluing triangulated categories" as it is defined in [**BBD82**, 1.4.3]—to which we also refer for terminology:

- $\underline{\mathrm{MCM}}(S)$ and $\underline{\mathrm{fpd}}(S)$ are full subcategories of $\underline{\mathrm{mod}}(S)$ (this corresponds to axiom 1.4.3.5. of (*loc. cit.*)),
- $\underline{\mathrm{MCM}}(S)$ is left-orthogonal to $\underline{\mathrm{fpd}}(S)$ and $\underline{\mathrm{fpd}}(S)$ is right-orthogonal to $\underline{\mathrm{MCM}}(S)$, (which is (5.1.1) (i) and (ii) rephrased and corresponds to axiom 1.4.3.3.),
- the embedding of $\underline{\mathrm{MCM}}(S)$ into $\underline{\mathrm{mod}}(S)$, (this corresponds to $j_!$), admits $\mathbf{M}$ as its right adjoint (j^* of axiom 1.4.3.2.),
- the embedding of $\underline{\mathrm{fpd}}(S)$ into $\underline{\mathrm{mod}}(S)$, (this corresponds to i_*), admits $\mathbf{H}$ as its left adjoint (i^* of axiom 1.4.3.1.),

and, most important

- the presentation in (5.3.1) describes the desired unique decomposition of an object in $\underline{\mathrm{mod}}(S)$ into its "components" in $\underline{\mathrm{MCM}}(S)$ and $\underline{\mathrm{fpd}}(S)$, (more precisely, (5.3.3) corresponds to the distinguished triangle $j_!j^* \to \mathrm{id} \to i_*i^* \to$ in 1.4.3.4.)

What is lacking from a "complete" gluing is essentially the existence of the other adjoints $(i^!, j_*)$: Even by adding projectives not any module can in general

[3]Compare with [**88**, Corollary 4.3].

be embedded into an MCM or represented as a quotient of a module of finite projective dimension by an MCM.

5.5 After this "meta-mathematical" digression, consider the following case(s) where an MCM-approximation can be neatly described:

PROPOSITION 5.5.1. *Let S be strongly Gorenstein, N an S-module such that $\operatorname{Ext}^i_S(N,S)=0$ except for a single value $i=n$. Then a(ny) MCM-approximation of N is given by*

$$\mathbf{M}(N)=\operatorname{Hom}_S(\Omega_n \operatorname{Ext}^n_S(N,S),S)\,,$$

31 *where $\Omega_n \operatorname{Ext}^n_S(N,S)$ denotes the corresponding syzygy module in some projective resolution of $\operatorname{Ext}^n_S(N,S)$ as left S-module.*

PROOF. The proof is a literal repetition of the proof of (5.1.2), observing that—with its notations—F^* is, by the special assumption, quasi-isomorphic to $\operatorname{Ext}^n_S(N,S)[-n]$, and hence G is a projective resolution of this module. But then $\mathbf{M}(N)$ is represented by

$$\operatorname{Ker}(\operatorname{Hom}_S(d^0_G,S))=\operatorname{Hom}_S(\operatorname{Cok}(d^0_G),S)=\operatorname{Hom}_S(\Omega_0 G,S)$$

and $\Omega_0 G=\Omega_n \operatorname{Ext}^n_S(N,S)$ by definition. □

Applying this to the case of a commutative local Gorenstein ring we get

COROLLARY 5.5.2. *Let R be a local commutative Gorenstein ring of dimension r with residue class field k. Then $\mathbf{M}(k)$ can be obtained as the R-dual of the r-th syzygy module in a (minimal) free resolution of k over R.* □

Sticking with the case of the Corollary, $\mathbf{M}(k)$ has a well-defined rank and choosing indeed a minimal resolution to obtain the MCM-approximation, its rank is given by

$$\operatorname{rank}_R \mathbf{M}(k)=\sum_{i=0}^{r-1}(-1)^{r-i-1}b_i\,, \tag{5.5.3}$$

where b_i denotes the i-th Betti number of k over R. Now, as in general the Betti numbers of a local Gorenstein ring grow rather fast, the rank of this maximal Cohen–Macaulay module will accordingly be quite large—and the ranks of its syzygy modules even larger, unless R is a hypersurface ring, (see [**Her78**], [**Eis80**]). Hence we ask:

Is the number given in (5.5.3) the minimal possible rank for a non-free MCM which occurs as the syzygy module of some artinian R-module?

Considering more generally the case of not necessarily commutative but local rings S which are strongly Gorenstein, one still has minimal resolutions, and an object in $\underline{\operatorname{mod}}(S)$ admits a unique (up to S-isomorphism) minimal representative in $\operatorname{mod}(S)$, characterized by the property that it does not contain a free summand. In particular, we can—and will—in this case "minimise" the functors $\mathbf{M}$ and $\mathbf{H}$ (as well as, for example, complete resolutions). This enables one to define several new invariants for any S-module N: The *rank* of $\mathbf{H}(N)$ and the *number of indecomposable summands* of $\mathbf{H}(N)$ or $\mathbf{M}(N)$.

32

5.6 The reader may have observed that the proof of (5.1.2) and (5.1.4), in particular (5.1.6), yields an extension of the construction of complete resolutions of MCMs, as given in (4.5), to arbitrary S-modules. Furthermore, essentially the same argument provides a *functorial* way to obtain such complete resolutions for arbitrary objects in $\mathrm{D^b}(S)$.

In other words, the projection functor $\mathrm{D^b}(S) \to \underline{\mathrm{D}}^{\mathrm{b}}(S)$ factors naturally over the equivalence $\underline{\sigma}_{\leqslant}\colon \underline{\mathrm{APC}}(S) \to \underline{\mathrm{D}}^{\mathrm{b}}(S)$, completing the picture given in (4.4).

Nevertheless the result is quicker to state than to formalise:

Complete resolutions can be obtained by "symmetrising" projective resolutions with respect to duality.

This shall mean:

- Assume given a complex X of S-modules—with bounded and finitely generated cohomology by our general assumption—and consider it as an object in $\mathrm{D^b}(S)$.
- Choose a projective resolution $\mathbf{P}(X) \to X$ of X and dualize it to obtain $\mathbf{P}(X)^* = \mathrm{Hom}_S(\mathbf{P}(X), S)$.
- By construction, this is a bounded below complex of finitely generated projective S^{op}-modules with bounded cohomology, (4.6), hence a "projective co-resolution" of $\mathrm{RHom}_S(X, S)$, considered as an object in the triangulated category $\mathrm{K}^{+,\mathrm{b}}(\mathrm{P}(S^{\mathrm{op}}))$. Therein, this object is unique up to homotopy equivalence.
- Now choose a projective resolution $\varphi_X\colon \mathbf{P}(\mathbf{P}(X)^*) \to \mathbf{P}(X)^*$ of this coresolution—it corresponds to the morphism φ of (5.1.6)—and dualize again, this time of course with respect to $\mathrm{Hom}_{S^{\mathrm{op}}}(-, S)$, to obtain
- the *Norm map* of X—called so here in analogy to the classical case of finite groups, see [**CE99**, XII.1] for example—:

$$\mathbf{N}(X)\colon \mathbf{P}(X) \xrightarrow[\cong]{\text{can.}} \mathbf{P}(X)^{**} \xrightarrow{\varphi_X^*} \mathbf{P}(\mathbf{P}(X)^*)^*, \tag{5.6.1}$$

  which by construction is a quasi-isomorphism from a projective resolution of X to a projective co-resolution of X. Considered in $\mathrm{D^b}(S)$, the morphism $\mathbf{N}(X)$ is hence nothing but the usual incarnation of the duality-isomorphism for Gorenstein rings:

$$X \xrightarrow{\simeq} \mathrm{RHom}_{S^{\mathrm{op}}}(\mathrm{RHom}_S(X, S), S)\,.$$

- Coming to an end, set—in accordance with (4.5)—:

$$\mathbf{CR}(X) := \mathrm{cone}(\mathbf{N}(X))[-1] \tag{5.6.2}$$

  and call this translated mapping cone over the Norm map a *complete resolution* 33
  of X.

This complete resolution of X gives, by its definition, rise to a distinguished triangle in $\mathrm{K}(\mathrm{P}(S))$, the homotopy category of *all* complexes of finitely generated projective (right) S-modules, cf. (0.3),:

$$\mathbf{CR}(X) \xrightarrow{\mathbf{r}(X)} \mathbf{P}(X) \xrightarrow{\mathbf{N}(X)} \mathbf{P}(\mathbf{P}(X)^*)^* \xrightarrow{\mathbf{i}(X)} \mathbf{CR}(X)[1]\,, \tag{5.6.3}$$

which is the typical distinguished triangle associated to the mapping cone over $\mathbf{N}(X)$ rotated once to the left. The occurring morphism $\mathbf{r}(X)$ is just the restriction of the complete resolution to the given projective resolution, whereas $\mathbf{i}(X)$ is the morphism "inflating" the chosen co-resolution to a complete resolution. (For the role of the shift, see once again the remark preceding (4.7) as well as (3.1).)

To abbreviate notations, we will set in the sequel:

$$\mathbf{C}(-) := \mathbf{P}(\mathbf{P}(-)^*)^* . \tag{5.6.4}$$

REMARKS. (a) Recall—[**Ver77**, II.1.4] dualized—that choosing projective resolutions just means categorically to give a left adjoint to the embedding

$$\mathrm{K}^{-,\mathrm{b}}(\mathrm{P}(S)) \to \mathrm{D}^{\mathrm{b}}(S) .$$

From this point of view, $\mathbf{C}$ becomes a functor from $\mathrm{D}^{\mathrm{b}}(S)$ into $\mathrm{K}^{+,\mathrm{b}}(\mathrm{P}(S))$, the homotopy category of projective co-resolutions. Composing each of these two functors with the respective inclusion $\mathrm{K}^{+/-,\mathrm{b}}(\mathrm{P}(S)) \to \mathrm{K}^{\mathrm{b}}(\mathrm{P}(S))$[4], the "norm" can hence the thought of as a natural transformation of functors from $\mathrm{D}^{\mathrm{b}}(S)$ into $\mathrm{K}^{\mathrm{b}}(\mathrm{P}(S))$. Similarly, (5.6.3) constitutes a functor into the category of (distinguished) triangles of functors, if one d/cares to introduce such a thing.

(b) As usual, any choice of a dualizing module ω_S will define a "norm" and then "complete resolutions" and so on, but all these objects will be isomorphic. Hence we will in general not distinguish these constructions and rather allow ourselves to choose the suitable dualizing module depending on the concrete situation.

To analyse further the distinguished triangle (5.6.3) we introduce the following two invariants of an object X in $\mathrm{D}^{\mathrm{b}}(S)$:

$$\begin{aligned} m_X &:= \max(i : \mathrm{H}^i(X) \neq 0) \qquad \text{and} \\ m_X^* &:= \max(i : \mathrm{H}^i(\mathrm{RHom}_S(X, S)) \neq 0) . \end{aligned} \tag{5.6.5}$$

(In terms of the t-structures mentioned in (4.8), these numbers measure how far the given object X is from the respective hearts of these structures: X is in the essential image of $\mathrm{MCM}(S)$ in $\mathrm{D}^{\mathrm{b}}(S)$ iff $m_X = m_X^* = 0$.)

Now choosing projective resolutions "minimally"—so that $\mathbf{P}^i(X) = 0$ for $i > m_X$—it follows that the morphisms in (5.6.3) can be assumed to satisfy:

- $\mathbf{r}(X)$ is the identity in each degree not larger than $-m_X^*$,
- $\mathbf{i}(X)$ is the identity in each degree not smaller than m_X,

and hence

- the Norm map $\mathbf{N}(X)$ is non-zero only in the "twilight-zone" of those degrees i for which $-m_X^* \leq i \leq m_X$.

Finally, having chosen complete resolutions for objects in $\mathrm{D}^{\mathrm{b}}(S)$, we define accordingly their *MCM-approximation*, extending (5.4.1) as

34 (5.6.6) $$\mathbf{M}(X) := \Omega_0 \mathbf{CR}(X) \quad \text{for any } X \text{ in } \mathrm{D}^{\mathrm{b}}(S) .$$

REMARK. (c) If compared to the situation of actual modules, what is missing in this more general context is an analogue of the functor $\mathbf{H}$, which would associate to any complex in $\mathrm{D}^{\mathrm{b}}(S)$ its "perfect part" in $\mathrm{D}^{\mathrm{b}}_{\mathrm{perf}}(S)$. But, even if such an analogue exists, there will be the same shortcoming as in (5.4): $\mathrm{D}^{\mathrm{b}}(S)$ can in general not be obtained by gluing together $\underline{\mathrm{D}}^{\mathrm{b}}(S)$ and $\mathrm{D}^{\mathrm{b}}_{\mathrm{perf}}(S)$ as simple examples already show—see also Section 7.[5]

---

[4]This denotes complexes of finitely generated projective modules with bounded cohomology.

[5]Such a gluing exists if one uses a large analogue of the stable derived category; see [**88**].

Remark that *if* there exists a stabilized higher K-theory as asked for in (4.9), an obstruction for the gluing will be given by any non-zero connecting homomorphism $\underline{\mathsf{K}}_i(S) \to \mathsf{K}_{i-1}(S)$.

To summarise the situation, we have seen:

THEOREM 5.6.7. *For any ring S which is strongly Gorenstein, one may extend* (4.4.1) *to a diagram of exact functors between triangulated categories, commutative up to isomorphism of such functors:*

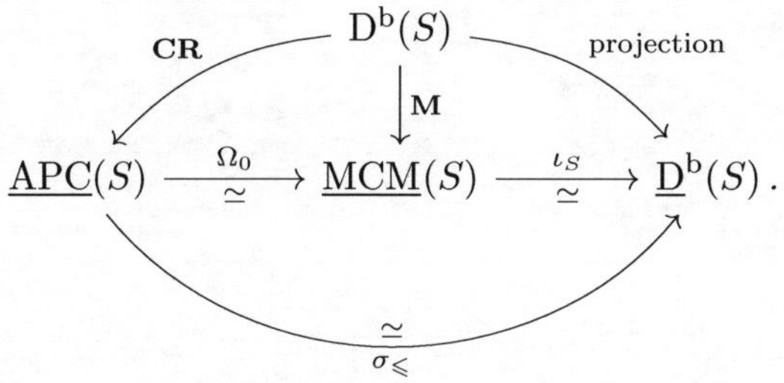

35

In particular, the functors **CR** *(resp.* **M***) induce quasi-inverses of $\sigma_{\leqslant}$ (resp. ι_S) and extend naturally the functors defined earlier,* (4.5.1) *(resp.* (5.4.1)*).* □

CHAPTER 6

The Tate cohomology

6.1 The essential use of the triangulated structures on any of the equivalent categories $\underline{\mathrm{APC}}(S)$, $\underline{\mathrm{MCM}}(S)$ or $\underline{\mathrm{D}}^{\mathrm{b}}(S)$ is that they define one and the same cohomology theory for S-modules (or, more generally, for complexes in $\mathrm{D}^{\mathrm{b}}(S)$):

DEFINITION 6.1.1. Let M, N be two (complexes of) modules (with bounded, finitely generated cohomology) over a ring S which is strongly Gorenstein. Then the *i-th Tate cohomology group of M with values in N* (over S) is defined to be

$$\underline{\mathrm{Ext}}^i_S(M,N) := \mathrm{Hom}_{\underline{\mathrm{D}}^{\mathrm{b}}(S)}(M, \mathrm{T}^i\, N)\,,$$

for any integer i.

A first more concrete description of these groups for actual S-modules can be obtained from Theorem 4.4.1 as follows:

LEMMA 6.1.2. *Let M and N be in* $\mathrm{mod}(S)$. *Then, for any i,*

(i) $\underline{\mathrm{Ext}}^i_S(M,N) = \underline{\mathrm{Hom}}_S(\Omega^i_S\mathbf{M}(M),N)$*—with notations as in Chapter 2—and also*
(ii) $\underline{\mathrm{Ext}}^i_S(M,N) = \mathrm{H}^i\,\mathrm{Hom}_S(\mathbf{CR}(M),N)$.

PROOF. By the defining properties of MCM-approximations, $\mathbf{M}(-)$ yields an **36**
inverse of ι_S, and by (4.4.1) (3) the loop-space functor Ω_S represents an inverse of the translation functor on $\underline{\mathrm{MCM}}(S)$, so that

$$\mathrm{Hom}_{\underline{\mathrm{D}}^{\mathrm{b}}(S)}(M,\mathrm{T}^i\,N) = \mathrm{Hom}_{\underline{\mathrm{D}}^{\mathrm{b}}(S)}(\mathrm{T}^{-i}\,M,N) = \mathrm{Hom}_{\underline{\mathrm{MCM}}(S)}(\Omega^i_S\mathbf{M}(M),\mathbf{M}(N))\,.$$

But, on the other hand, (5.4), (the restriction of) $\mathbf{M}$ on $\underline{\mathrm{mod}}(S)$ is a right adjoint of the embedding of $\underline{\mathrm{MCM}}(S)$ into $\underline{\mathrm{mod}}(S)$, so that

$$\mathrm{Hom}_{\underline{\mathrm{MCM}}(S)}(\Omega^i_S\mathbf{M}(M),\mathbf{M}(N)) = \mathrm{Hom}_{\underline{\mathrm{mod}}(S)}(\Omega^i_S\mathbf{M}(M),N)\,,$$

whence (i).

To verify (ii), remark that $\mathbf{CR}(-)$ establishes an inverse of $\sigma_{\leqslant}$, by (5.6.7), and that by (4.4.1) (v)[1] one may replace $\sigma_{\leqslant}$ by $\sigma_{\leqslant k}$ for some sufficiently large k—depending on i—so that the cohomology group in question can be calculated as:

$$\mathrm{Hom}_{\underline{\mathrm{D}}^{\mathrm{b}}(S)}(M,\mathrm{T}^i\,N) = \mathrm{Hom}_{\mathrm{K}(S)}(\sigma_{\leqslant k}\mathbf{CR}(M),\mathrm{T}^i\,N)$$

which last group of homotopy classes of morphisms of complexes equals by definition $\mathrm{H}^i\,\mathrm{Hom}_S(\sigma_{\leqslant k}\mathbf{CR}(M),N)$—see [**Bou80**, p.82] for example.

Now for k large enough (again compared to i), "$\sigma_{\leqslant k}$" can obviously be dropped as the "complex" N is bounded above. □

37

[1] For clarification see A.6.1.

6.2 Before we give a more classical description of the newly introduced cohomology theory, we use the general machinery—in particular (4.4.1) and (5.6)—to investigate how the ordinary Ext compares with the stabilized $\underline{\mathrm{Ext}}$.

This will be done by calculating the Tate cohomology as homotopy-groups in $\underline{\mathrm{APC}}(S)$, considered as a full triangulated subcategory of $\mathrm{K}(\mathrm{P}(S))$, the homotopy category of all complexes of finitely generated projective S-modules.

For this purpose and the reader's convenience, let us recall the following facts on homotopy-groups of such complexes:

LEMMA 6.2.1. *Let S be any ring, X and Y two complexes of finitely generated projective (right) S-modules. Denote—as before—by* $\mathrm{Hom}_S(X,Y)$ *the complex of morphisms of the underlying graded S-modules, (which, following* [**Bou80**, p.81], *should be* $\mathrm{Homgr}_S(X,Y)$*).*

(i) $\mathrm{Hom}_S(X,Y)$ *is acyclic if*
   (a) *X is bounded above and Y is acyclic, or*
   (b) *the injective dimension of S as a right module over itself is finite, X is acyclic and Y is bounded below.*

(ii) *The natural morphism of complexes* $Y \otimes_S \mathrm{Hom}_S(X,S) \to \mathrm{Hom}_S(X,Y)$*—as defined in* [**Bou80**, p.99 (16)] *say—is an isomorphism if X is bounded below and Y is bounded above.*

PROOF. (a) of (i) is well-known—dualize [**Har66**, I.4.4.] e.g.—and (ii) is obvious from the definitions of the complexes involved. Statement (b) in (i) can be seen most easily by extending the argument in (v) of the proof of (4.4.1).[2] □

This lemma yields the following information on homotopy classes of morphisms between the various functor **CR**, **P** or **C** as introduced in (5.6):

LEMMA 6.2.2. *Let S be a ring which is strongly Gorenstein, X and Y two complexes in* $\mathrm{D^b}(S)$.

(i) *The complex of abelian groups* $\mathrm{Hom}_S(\mathbf{CR}(X), \mathbf{C}(Y))$ *is acyclic and the restriction-morphism* $\mathbf{r}(Y)\colon \mathbf{CR}(Y) \to \mathbf{P}(Y)$ *of* (5.6.3) *induces a quasi-isomorphism*

$$\mathrm{Hom}_S(\mathbf{CR}(X), \mathbf{r}(Y))\colon \mathrm{Hom}_S(\mathbf{CR}(X), \mathbf{CR}(Y)) \xrightarrow{\simeq} \mathrm{Hom}_S(\mathbf{CR}(X), \mathbf{P}(Y))\,.$$

(ii) *The complex of abelian groups* $\mathrm{Hom}_S(\mathbf{P}(X), \mathbf{CR}(Y))$ *is acyclic and the Norm map of Y,* $\mathbf{N}(Y)$ *induces a quasi-isomorphism*

$$\mathrm{Hom}_S(\mathbf{P}(X), \mathbf{N}(Y))\colon \mathrm{Hom}_S(\mathbf{P}(X), \mathbf{P}(Y)) \xrightarrow{\simeq} \mathrm{Hom}_S(\mathbf{P}(X), \mathbf{C}(Y))\,.$$

38 PROOF. The first statement in (i) is case (b) and the first statement in (ii) is case (a) of the foregoing lemma. The remaining assertions follow then by applying $\mathrm{Hom}_S(\mathbf{CR}(X), -)$ resp. $\mathrm{Hom}_S(\mathbf{P}(X), -)$ to the distinguished triangle (5.6.3) for Y. □

To state the main result in Tate cohomology we use—as in (5.6)—the

NOTATION 6.2.3. If X is any complex of finitely generated projective S-modules (any object of $\mathrm{D^b}(S)$), we denote X^* the complex of finitely generated projective S^{op}-modules $\mathrm{Hom}_S(X,S)$, (resp. the object $\mathrm{RHom}_S(X,S)$ in $\mathrm{D^b}(S^{\mathrm{op}})$).

[2] For clarification see A.6.2.

Accordingly we denote for two objects X and Y in $\mathrm{D^b}(S)$ by

$$\mathrm{Tor}_i^S(Y, X^*) = \mathrm{H}^{-i}(Y \otimes_S^{\mathsf{L}} \mathrm{RHom}_S(X, S))$$

the corresponding *hyper-Tor* group, which is certainly defined as soon as S is of finite injective dimension as a right module over itself.[3]

Remark that by the very definition, these hyper-Tor groups can be calculated as $\mathrm{Tor}_i^S(Y, X^*) = \mathrm{H}^{-i}(\mathbf{P}(Y) \otimes_S \mathbf{P}(\mathrm{Hom}_S(\mathbf{P}(X), S)))$, hence using (5.6.4) and (6.2.1) (ii) one obtains for any two complexes X, Y in $\mathrm{D^b}(S)$ with S strongly Gorenstein:

$$\mathrm{Tor}_i^S(Y, X^*) = \mathrm{H}^{-i}\,\mathrm{Hom}_S(\mathbf{C}(X), \mathbf{P}(Y))\,. \tag{6.2.4}$$

With these preparations—and the notations in (5.6)—we get:

THEOREM 6.2.5. *Let S be a ring which is strongly Gorenstein.*

(1) *For any two complexes X and Y in $\mathrm{D^b}(S)$, the complex of abelian groups $\mathrm{Hom}_S(\mathbf{CR}(X), \mathbf{CR}(Y))$ is quasi-isomorphic to the mapping cone over*

$$\mathbf{N}(X, Y) := \mathrm{Hom}_S(\mathbf{N}(X), \mathbf{N}(Y))\,.$$

More precisely, there is a quasi-isomorphism $\mathbf{d}(X, Y)$, so that the upper left triangle in the following diagram is commutative:

$$\begin{array}{ccc}
\mathrm{Hom}_S(\mathbf{CR}(X)[1], \mathbf{CR}(Y))[1] = \mathrm{Hom}_S(\mathbf{CR}(X), \mathbf{CR}(Y)) & \xrightarrow{\mathbf{d}(X,Y)} & \mathrm{cone}(\mathbf{N}(X, Y)) \\
{\scriptstyle \mathrm{Hom}_S(\mathbf{i}(X), \mathbf{r}(Y))[1]}\downarrow{\scriptstyle (1)} \qquad \swarrow{\scriptstyle d} \quad (1) \quad (+) & & \uparrow{\scriptstyle j} \\
\mathrm{Hom}_S(\mathbf{C}(X), \mathbf{P}(Y)) & \xrightarrow[\mathbf{N}(X,Y)]{} & \mathrm{Hom}_S(\mathbf{P}(X), \mathbf{C}(Y))\,.
\end{array}$$

The triangle marked $(+)$ is a "typical" distinguished triangle associated to the mapping cone over $\mathbf{N}(X, Y)$. 39

(2) *The above diagram is self-dual in the sense that "transposition" with respect to $(-)^* = \mathrm{Hom}_{S^{(\mathrm{op})}}(-, S)$ yields isomorphisms of distinguished triangles in the (full) derived category of abelian groups*

$$\begin{array}{ccccc}
\mathrm{Hom}_S(\mathbf{C}(X), \mathbf{P}(Y)) & \xrightarrow{\mathbf{N}} & \mathrm{Hom}_S(\mathbf{P}(X), \mathbf{C}(Y)) & \xrightarrow{\mathbf{d}^{-1}j} & \mathrm{Hom}_S(\mathbf{CR}(X), \mathbf{CR}(Y)) \\
\simeq\updownarrow{\scriptstyle (-)^*} & & \simeq\updownarrow{\scriptstyle (-)^*} & & \simeq\updownarrow{\scriptstyle (-)^*[-1]} \\
\mathrm{Hom}_{S^{\mathrm{op}}}(\mathbf{C}(Y^*), \mathbf{P}(X^*)) & \xrightarrow{\mathbf{N}} & \mathrm{Hom}_{S^{\mathrm{op}}}(\mathbf{P}(Y^*), \mathbf{C}(X^*)) & \xrightarrow{\mathbf{d}^{-1}j} & \mathrm{Hom}_{S^{\mathrm{op}}}(\mathbf{CR}(Y^*), \mathbf{CR}(X^*))
\end{array}$$

for any two objects X and Y in $\mathrm{D^b}(S)$. Here, $\mathbf{d}^{-1}$ represents of course the inverse of $\mathbf{d}$ in the derived category.

(3) *Evaluating 1 and 2 by taking cohomology, one obtains a long—even unlimited—exact sequence of abelian groups*

$$\longrightarrow \underline{\mathrm{Ext}}_S^{i-1}(X, Y) \longrightarrow \underline{\mathrm{Tor}}_{-i}^S(Y, X^*) \xrightarrow{\mathrm{H}^i\,\mathbf{N}(X,Y)} \mathrm{Ext}_S^i(X, Y) \xrightarrow{c^i} \underline{\mathrm{Ext}}_S^i(X, Y) \longrightarrow$$

where c^i is the map obtained by applying the projection functor

$$\mathrm{Ext}_S^i(X, Y) = \mathrm{Hom}_{\mathrm{D^b}(S)}(X, \mathrm{T}^i\,Y) \longrightarrow \mathrm{Hom}_{\underline{\mathrm{D}}^{\mathrm{b}}(S)}(X, \mathrm{T}^i\,Y) = \underline{\mathrm{Ext}}_S^i(X, Y)\,,$$

as well as isomorphisms

$$\underline{\mathrm{Ext}}_S^i(X, Y) \cong \underline{\mathrm{Ext}}_{S^{\mathrm{op}}}^i(Y^*, X^*)$$

for all integers i and all objects X, Y in $\mathrm{D^b}(S)$.

---

[3] In fact, the hyper-Tor groups are always defined; no boundedness assumptions are needed to construct the derived tensor product.

(4) *All statements as well as diagrams in* 1–3 *are functorial in* X *and* Y.

PROOF. 1 Remark first that the *existence* of the claimed morphism $\mathbf{d}(X,Y)$ follows readily from the fact that

$$\mathbf{N}(X,Y)\cdot \mathrm{Hom}_S(\mathbf{i}(X),\mathbf{r}(Y)) = \mathrm{Hom}_S(\mathbf{i}(X)\mathbf{N}(X),\mathbf{N}(Y)\mathbf{r}(Y)) = 0$$

by definition of the morphisms involved, (5.6.3).

That $\mathbf{d}(X,Y)$ is necessarily a *quasi-isomorphism* follows from the so-called
40 "diagram of 9 squares"—see [**BBD82**, 1.1.11.]—obtained by applying $\mathrm{Hom}_S(-,-)$ to the distinguished triangle (5.6.3) for X in its contravariant argument and to the corresponding triangle for Y in its covariant argument.

In the following display, we abbreviate $\mathrm{Hom}_S(-,-)$ as $(-,-)$ and suppress the arguments X and Y. The rows and columns in solid arrows represent distinguished triangles, the right-most column (bottom row) are the "anti-distinguished" triangles obtained (up to canonical isomorphism) by translating the left-most column (the top row).

The square marked $(-)$ is anti-commutative.

(6.2.6)

$$\begin{array}{ccccccc}
(\mathbf{CR},\mathbf{CR}) & \xrightarrow{\simeq} & (\mathbf{CR},\mathbf{P}) & \longrightarrow & \boxed{(\mathbf{CR},\mathbf{C})} & \longrightarrow & (\mathbf{CR},\mathbf{CR})[1] \\
\uparrow & & \uparrow & & \uparrow & & \uparrow \\
\boxed{(\mathbf{P},\mathbf{CR})} & \longrightarrow & (\mathbf{P},\mathbf{P}) & \xrightarrow{\simeq} & (\mathbf{P},\mathbf{C}) & \longrightarrow & \boxed{(\mathbf{P},\mathbf{CR})[1]} \\
\uparrow & & \uparrow & (\#) \quad \mathbf{N}(X,Y) & \uparrow\simeq & & \uparrow \\
(\mathbf{C},\mathbf{CR}) & \longrightarrow & (\mathbf{C},\mathbf{P}) & \longrightarrow & (\mathbf{C},\mathbf{C}) & \longrightarrow & (\mathbf{C},\mathbf{CR})[1] \\
\simeq\uparrow & (\mathbf{i},\mathbf{r}) & \uparrow & & \uparrow & (-) & \simeq\uparrow \\
(\mathbf{CR}[1],\mathbf{CR}) & \xrightarrow{\simeq} & (\mathbf{CR}[1],\mathbf{P}) & \longrightarrow & \boxed{(\mathbf{CR}[1],\mathbf{C})} & \longrightarrow & (\mathbf{CR}[1],\mathbf{CR})[1]
\end{array}$$

In this diagram, by (6.2.2), the boxed complexes are acyclic and then the morphisms marked "$\simeq$" are quasi-isomorphisms.

The factorisation—marked by $(\#)$ above—of $\mathbf{N}(X,Y) = \mathrm{Hom}_S(\mathbf{N}(X),\mathbf{N}(Y))$ over the quasi-isomorphism $\mathrm{Hom}_S(\mathbf{P}(X),\mathbf{N}(Y))$ implies the existence of a morphism from $\mathrm{Hom}_S(\mathbf{CR}(X),\mathbf{P}(Y))$ into the mapping cone over $\mathbf{N}(X,Y)$ which is necessarily a quasi-isomorphism. Now the claimed morphism $\mathbf{d}(X,Y)$ can be obtained by composing this with the quasi-isomorphism

$$\mathrm{Hom}_S(\mathbf{CR}(X),\mathbf{CR}(Y)) \xrightarrow{\simeq} \mathrm{Hom}_S(\mathbf{CR}(X),\mathbf{P}(Y)),$$

the horizontal arrow in the upper left corner, whence $\mathbf{d}(X,Y)$ itself is also a quasi-isomorphism.

Remark that this proof exhibits isomorphisms in the derived category (of abelian groups) between the distinguished triangles represented respectively by the
41 second column or the third row and the triangle given by $(\mathbf{N}(X,Y),\mathbf{d}^{-1}j,(\mathbf{i},\mathbf{r})[1])$.

Strictly speaking, this last triple does not represent a distinguished triangle, unless the morphism $\mathrm{Hom}_S(\mathbf{i}(X),\mathbf{r}(Y))[1]$, whose source is by definition

$$\mathrm{Hom}_S(\mathbf{CR}(X)[1],\mathbf{CR}(Y))[1],$$

is composed with the *canonical*—[**AGV72**, XVII.1.1.5.]—isomorphism from

$$\mathrm{Hom}_S(\mathbf{CR}(X), \mathbf{CR}(Y))$$

into its source, as indicated in the diagram of 1.

Taking into account the foregoing remark, it is clear from the definitions that "transposition" corresponds to "reflecting" the above diagram on the "diagonal".

For example

$$\mathbf{C}(X)^* = \mathbf{P}(\mathbf{P}(X)^*)^{**} \cong \mathbf{P}(X^*)$$

and

$$\mathbf{P}(Y)^* \cong \mathbf{P}(Y^{**})^* = \mathbf{P}(\mathbf{P}(Y^*)^*)^* = \mathbf{C}(Y^*)$$

by (6.2.3), (5.6.3), and the duality on complete resolutions which has been observed already in (4.6.1).

That these dualities are in fact compatible, so that the given diagram in 2 commutes, is left to the reader.

Concerning 3, recall that always $\mathrm{Ext}^i_S(X,Y) = \mathrm{H}^i \mathrm{Hom}_S(\mathbf{P}(X), \mathbf{P}(Y))$, so that (6.2.4) and passing to cohomology in (6.2.2) (ii) and (6.2.6) yields the identifications needed. (See also Remark (b) below.)

The functoriality asserted in 4 is evident. □

REMARKS. (a) Obviously, the theorem contains an understatement: As usual, one may replace throughout its formulation—or proof—the term "abelian group" by "modules over the center of S".

(b) The proof of the theorem—in particular (6.2.6)—reveals the following possibilities to obtain in practice the natural transformation c: $\mathrm{Ext}_S(-,-) \to \underline{\mathrm{Ext}}_S(-,-)$: Identify these groups with the cohomology of

- $\mathrm{Hom}_S(\mathbf{P},\mathbf{P})$, $\mathrm{Hom}_S(\mathbf{P},\mathbf{C})$ or $\mathrm{Hom}_S(\mathbf{C},\mathbf{C})$ for Ext_S—this corresponds to the "central square" of (6.2.6)—and accordingly with
- $\mathrm{Hom}_S(\mathbf{CR},\mathbf{P})$, $\mathrm{Hom}_S(\mathbf{CR},\mathbf{CR})$ or $\mathrm{Hom}_S(\mathbf{C},\mathbf{CR}[1])$ for $\underline{\mathrm{Ext}}_S$ so that c becomes represented—in the corresponding order—by
- $\mathrm{Hom}_S(\mathbf{r},\mathbf{P})$, $\mathbf{d}^{-1}j$ or $\mathrm{Hom}_S(\mathbf{C},\mathbf{i})$.

Furthermore, in the first of these incarnations, one may drop the "**P**" in the covariant argument and take the complex itself instead—using once again (4.4.1) (v) and (6.2.1) (a).

The first of these descriptions is closest to a "classical" treatment, whence we 42
make it explicit—it has been used implicitly already, e.g. in (4.4.1) (v), (6.1.2):

- Assume given two (complexes of) S-modules X, Y (with bounded and finitely generated cohomology).
- Choose a projective resolution $\mathbf{P}(X)$ and construct from it a complete resolution $\mathbf{CR}(X)$ as in (5.6) and (4.5).
- The homotopy classes of morphisms of complexes of degree i from $\mathbf{P}(X)$, (resp. $\mathbf{CR}(X)$), into Y represent the i-th (stabilized) cohomology group of X with values in Y.
- Representing a homotopy class by an actual morphism of complexes from $\mathbf{P}(X)$ to Y, composing it with the restriction morphism $\mathbf{r}(X)$ from $\mathbf{CR}(X)$ onto $\mathbf{P}(X)$ and passing again to the homotopy class describes the image of an element from $\mathrm{Ext}^i_S(X,Y)$ under c^i in $\underline{\mathrm{Ext}}^i_S(X,Y)$.

The above theorem is then just the triangulated version of saying that this procedure is independent of the choices involved—and furthermore identifies the mapping cone over these natural maps c^i.

6.3 As observed in (5.6), for a given object X in $\mathrm{D^b}(S)$ the Norm map $\mathbf{N}(X)$ can be chosen to be non-trivial in only finitely many degrees. This implies that the exact sequence in (6.2.5) 3 is quite "degenerate".

More precisely, in addition to the invariants m_X and m_X^* introduced in (5.6.5), consider also

$$\begin{aligned} i_X &:= \inf(i : \mathrm{H}^i(X) \neq 0) \qquad \text{and} \\ i_X^* &:= \inf(i : \mathrm{H}^i(\mathrm{RHom}_S(X,S)) \neq 0)\,. \end{aligned} \tag{6.3.1}$$

In terms of these invariants, one gets immediately

LEMMA 6.3.2. *With notations as above, for any given complexes X and Y in $\mathrm{D^b}(S)$, one has:*

(i) $\mathrm{Tor}^S_{-i}(Y, X^*) = 0$ *for all* $i > m_X^* + m_Y$, *and*
(ii) $\mathrm{Ext}^i_S(X,Y) = 0$ *for all* $i < \max(i_Y - m_X, i_X^* - m_Y^*)$.

PROOF. By (6.2.4), $\mathrm{Tor}^S_{-i}(Y, X^*) = \mathrm{H}^i(\mathbf{P}(Y) \otimes_S \mathbf{P}(\mathbf{P}(X)^*))$, and the complex
43 $\mathbf{P}(Y) \otimes_S \mathbf{P}(\mathbf{P}(X)^*)$ can be assumed to be zero in degrees larger than $m_Y + m_X^*$ by definition of these numbers, (5.6.5), whence (i). For (ii), remark that by duality it is enough to have the claimed vanishing for all $i < i_Y - m_X$, which is obvious. □

In view of this lemma, we set

$$\begin{aligned} \underline{m} &:= \underline{m}(X,Y) = \max(i_Y - m_X, i_X^* - m_Y^*) - 1 \quad \text{and} \\ \overline{m} &:= \overline{m}(X,Y) = m_X^* + m_Y \end{aligned} \tag{6.3.3}$$

to obtain

COROLLARY 6.3.4. *Using the just defined integers, depending on X and Y in $\mathrm{D^b}(S)$, the long exact sequence of* (6.2.5) (3) *breaks up according to*

(i) *The natural homomorphism* $c^i(X,Y)\colon \mathrm{Ext}^i_S(X,Y) \to \underline{\mathrm{Ext}}^i_S(X,Y)$ *is*
  - *an isomorphism for* $i > \overline{m}$ *and*
  - *a surjection for* $i = \overline{m}$.

(ii) *The connecting homomorphism* $\underline{\mathrm{Ext}}^i_S(X,Y) \to \mathrm{Tor}^S_{-i-1}(Y, X^*)$ *is*
  - *an isomorphism for* $i < \underline{m}$ *and*
  - *an injection for* $i = \underline{m}$.

(iii) *It remains an exact sequence*

$$\begin{aligned} 0 \longrightarrow \underline{\mathrm{Ext}}^{\underline{m}}_S(X,Y) \longrightarrow \mathrm{Tor}^S_{-\underline{m}-1}(Y,X^*) \longrightarrow \cdots \longrightarrow \mathrm{Ext}^{\overline{m}}_S(X,Y) \\ \longrightarrow \underline{\mathrm{Ext}}^{\overline{m}}_S(X,Y) \longrightarrow 0\,. \end{aligned} \tag{6.3.5}$$

REMARK. The converging spectral sequence

$$E_2^{i,j} = \mathrm{Ext}^i_S(\mathrm{H}^{-j}(X), S) \implies \mathrm{H}^{i+j}(\mathrm{RHom}_S(X,S))$$

for a complex X in $\mathrm{D^b}(S)$ shows that always

$$-m_X \le i_X^* \le m_X^* \le \operatorname{vdim} S - i_X\,,$$

which implies readily that for any two complexes X and Y as above necessarily $\underline{m} < \overline{m}$, so that the exact sequence in (iii) above is in any case non-trivial. (Easy examples show furthermore that in general the bounds given in (6.3.2) cannot be improved, so that in this sense the above corollary is best possible.)

EXAMPLE 6.3.6. Let M, N be two S-modules, S strongly Gorenstein. Then

$$\underline{m} = \underline{m}(M,N) = \max(0, i_M^* - m_N^*) - 1 \geq -1\,,$$
$$0 \leq \overline{m} = \overline{m}(M,N) = m_M^* \leq \operatorname{vdim} S\,.$$

So one gets the universal bounds—improving (6.3.4)— **44**

$$\underline{\operatorname{Ext}}_S^i(M,N) = \operatorname{Ext}_S^i(M,N) \qquad \text{for all} \quad i > \operatorname{vdim} S \geq m_M^*\,,$$
$$\underline{\operatorname{Ext}}_S^i(M,N) = \operatorname{Tor}_{-i-1}^S(N, \operatorname{RHom}_S(M,S)) \quad \text{for all} \quad i < -1 \quad \leq \underline{m}\,.$$

If for any S-module N we call—in accordance with the local commutative case—

m_N^* the *codepth* of N, and recall that by definition
i_N^* is the *grade* of N, [**AB69**, 2.21.],

the above can be rephrased as

For S-modules M, N, the Tate cohomology of M with values in N yields "new" (stable) groups $\underline{\operatorname{Ext}}_S^i(M,N)$ only in the range

$$-1 \leq \max(0, \operatorname{grade}(M) - \operatorname{codepth}(N)) - 1 \leq i \leq \operatorname{codepth}(M) \leq \operatorname{vdim} S\,.$$

6.4 It follows from the example—and could be expected in view of (4.4.1) and the definition—that Tate cohomology takes a particularly simple form on maximal Cohen–Macaulay modules.

COROLLARY 6.4.1. *Let M be a maximal Cohen–Macaulay module over S and denote $M^* = \operatorname{Hom}_S(M,S)$ its S-dual. Then*

(i) $\underline{\operatorname{Ext}}_S^i(M,-) = \operatorname{Ext}_S^i(M,-)$ *for all $i > 0$,*
(ii) $\underline{\operatorname{Ext}}_S^i(M,-) = \operatorname{Tor}_{-i-1}^S(-,M^*)$ *for all $i < -1$,*
(iii) *there is an exact sequence of functors*

$$0 \longrightarrow \underline{\operatorname{Ext}}_S^{-1}(M,-) \longrightarrow - \otimes_S M^* \xrightarrow{\mathbf{n}(M,-)} \operatorname{Hom}_S(M,-) \longrightarrow \underline{\operatorname{Ext}}_S^0(M,-) \longrightarrow 0\,,$$

where for any (right) S-module N the morphism $\mathbf{n}(M,N)$ is given by

$$(\mathbf{n}(M,N)(n \otimes \phi))(m) = n \cdot \phi(m)$$

for any $n \in N$, $\phi \in M^*$ and $m \in M$, and
(iv) $\underline{\operatorname{Ext}}_S^0(M,N) = \underline{\operatorname{Hom}}_S(M,N)$, *the S-linear homomorphisms from M to N modulo those which factor over projective modules.*

PROOF. The first two statements are obtained from the above example and the foregoing corollary as by definition M is MCM iff

$$\operatorname{codepth}(M) = 0 = \operatorname{grade}(M)\,.$$

(iv) is just (6.1.2) (i) reformulated for $i = 0$.

Instead of deducing (iii) directly from the above results, we make the following remark. □

REMARK. In more generality, M. Auslander—[**67**, Ch.3, Prop.1 bis], see also **45**
[**AB69**, 2.8.]—introduced and studied the exact sequence of functors

(6.4.2)

$$0 \longrightarrow \operatorname{Tor}_2^T(-, D(M)) \longrightarrow - \otimes_T M^* \xrightarrow{\mathbf{n}(M,-)} \operatorname{Hom}_T(M,-) \longrightarrow \operatorname{Tor}_1^T(-, D(M)) \longrightarrow 0$$

on $\underline{\operatorname{mod}}(T)$, where

- T is an arbitrary ring,
- M is a (right) T-module which admits a presentation

$$Q \xrightarrow{f} P \longrightarrow M \longrightarrow 0$$

by finitely generated projective T-modules P and Q, and
- $D(M)$ is the left T-module obtained by dualizing the presentation with respect to T:

$$0 \longrightarrow M^* \longrightarrow P^* \xrightarrow{f^*} Q^* \longrightarrow D(M) \longrightarrow 0\,.$$

$D(M)$ is well-defined in the category of left T-modules modulo projectives.[4]

If now $T = S$ is strongly Gorenstein, M a maximal Cohen–Macaulay S-module, then

- $D(M) = \ker(f)^* = (\Omega_S^2 M)^*$ and $\Omega_S^2 M$ is again MCM, (4.2.2) (ii), so that
- $\mathrm{Tor}_i^S(-, D(M)) = \mathrm{Tor}_i^S(-, (\Omega_S^2 M)^*) = \underline{\mathrm{Ext}}_S^{-i-1}(\Omega_S^2 M, -)$ for all $i > 0$ by assertion (6.4.1) (ii) above. But, in view of (6.1.2) (i) together with the fact that Ω_S on $\underline{\mathrm{MCM}}(S)$ represents the inverse of the translation functor, this shows
- $\mathrm{Tor}_i^S(-, D(M)) = \underline{\mathrm{Ext}}_S^{-i+1}(M, -)$ for all $i > 0$,

whence the exact sequence (6.4.1) (iii) becomes a special case of the exact sequence above.

One may take (6.4.1) as the starting point for a "synthetic" definition of Tate cohomology by "dimension shifts":

For any ring S which is strongly Gorenstein, there is a unique family of cohomological bi-functors $\underline{\mathrm{Ext}}_S(-,-)$, characterised by

(i) (*δ-functor*) In classical terminology, $\underline{\mathrm{Ext}}_S(-,-)$ is a δ-functor in both arguments, contravariant in the first, covariant in the second.

(ii) (*effaceable and co-effaceable*) For any module U of finite projective dimension and for any integer i, one has

46 $$\underline{\mathrm{Ext}}_S^i(U, -) = 0 = \underline{\mathrm{Ext}}_S^i(-, U)\,,$$

and for any finitely generated S-module N there exist modules U and V of finite projective dimension, such that N is a quotient of U—obvious—and a submodule of V—by (5.1.4).

(iii) (*initial condition*) For a maximal Cohen–Macaulay S-module in the contravariant argument, these cohomological functors are determined by (6.4.1).

It is clear, that these three axioms determine the Tate cohomology groups completely and in a unique way for all modules in $\mathrm{mod}(S)$.

6.5 As a final observation in this direction, we will use the above description to "narrow the gap" in the "unstable range", (6.3.6), between -1 and $\operatorname{vdim} S$.

For this, remark first that the very historical reason for the introduction of the loop-space functor Ω_S on $\underline{\mathrm{mod}}(S)$ was the isomorphism of functors

$$\mathrm{Ext}_S^i(\Omega_S^j -, -) \xrightarrow{\cong} \mathrm{Ext}_S^{i+j}(-,-)$$

for all $i > 0$, $j \geq 0$—cf. [**Bou80**, 128; Cor.4].

[4] This is the Auslander transpose of M.

On the other hand, by definition of the groups involved, there are natural surjections

$$\mathrm{Ext}^i_S(-,-) \to \underline{\mathrm{Hom}}_S(\Omega^i_S -,-)$$

for all $i \geq 0$. These surjections yield precisely the image of the natural transformation

$$c^i\colon \mathrm{Ext}^i_S(-,-) \to \underline{\mathrm{Ext}}^i_S(-,-)$$

in case of a ring S which is strongly Gorenstein. More specifically:

PROPOSITION 6.5.1. *For any two modules M, N in* $\mathrm{mod}(S)$ *one has natural isomorphisms*

$$\varinjlim_n \underline{\mathrm{Hom}}_S(\Omega^{i+n}_S M, \Omega^n_S N) \xrightarrow{\cong} \underline{\mathrm{Ext}}^i_S(M,N)$$

for all integers i.

The direct limit on the left is essentially constant in the sense that all transition maps are isomorphisms as soon as $i+n$ is at least equal to $\mathrm{vdim}\, S$. *In particular, this limit "collapses" if $i \geq$* $\mathrm{vdim}\, S$, *but defines a finite ascending filtration of* $\underline{\mathrm{Ext}}^i_S(M,N)$ *by abelian subgroups*

$$F^i_k = \mathrm{Im}(\underline{\mathrm{Hom}}_S(\Omega^{i+k}_S M, \Omega^k_S N) \longrightarrow \underline{\mathrm{Ext}}^i_S(M,N)), \quad k \geq \max(0,-i)\,.$$

For $i \geq 0$, F^i_k equals the image of $\mathrm{Ext}^i_S(\Omega^k_S M, \Omega^k_S N)$ *in* $\underline{\mathrm{Ext}}^i_S(M,N)$.

PROOF. This statement can be obtained directly from the axiomatic treatment 47
above, or, more conceptually, by applying the functor of MCM-approximations:

As $\mathbf{M}(-)$ is a right adjoint of the embedding $\underline{\mathrm{MCM}}(S) \to \underline{\mathrm{mod}}(S)$, it commutes necessarily with the loop-space functor. But as Ω_S on $\underline{\mathrm{MCM}}(S)$ represents the inverse of the translation functor, one gets

$$\underline{\mathrm{Hom}}_S(\Omega^{i+n}_S M, \Omega^n_S N) \longrightarrow \mathrm{Hom}_{\underline{\mathrm{MCM}}(S)}(\mathbf{M}(\Omega^{i+n}_S M), \mathbf{M}(\Omega^n_S N))$$

by applying $\mathbf{M}$ and

$$\begin{aligned}
\mathrm{Hom}_{\underline{\mathrm{MCM}}(S)}(\mathbf{M}(\Omega^{i+n}_S M), \mathbf{M}(\Omega^n_S N)) &= \mathrm{Hom}_{\underline{\mathrm{MCM}}(S)}(\Omega^{i+n}_S \mathbf{M}(M), \Omega^n_S \mathbf{M}(N))\\
&= \mathrm{Hom}_{\underline{\mathrm{MCM}}(S)}(\mathrm{T}^n\, \mathbf{M}(M), \mathrm{T}^{i+n}\, \mathbf{M}(N))\\
&= \mathrm{Hom}_{\underline{\mathrm{MCM}}(S)}(\mathbf{M}(M), \mathrm{T}^i\, \mathbf{M}(N))\\
&= \underline{\mathrm{Ext}}^i_S(M,N)
\end{aligned}$$

using the definition (6.1.1) and the equivalence ι_S of (4.4.1) (2). Applying Ω_S and passing to the limit establishes then the result in the view of the fact that one has an isomorphism of functors $\mathbf{M}\Omega^j_S = \Omega^j_S$ as soon as $j \geq \mathrm{vdim}\, S$, cf. (4.2.2) (iv). □

This proposition shows in which sense Tate cohomology stabilizes the ordinary Ext's and provides the link to the "stable module theory" of M. Auslander–M. Bridger, [**AB69**]. Furthermore, it reveals a fourth equivalent description of $\underline{\mathrm{MCM}}(S)$, the category of all maximal Cohen–Macaulay modules modulo projectives over a ring which is strongly Gorenstein:

> *It can be obtained—up to natural equivalence—from* $\underline{\mathrm{mod}}(S)$, *the stable module category, by "inverting" the loop-space functor Ω_S.*

To formalise this statement, remark first the following general fact:

LEMMA 6.5.2. *Assume given a pair* (A, e)*, where* A *is an additive category,* $e\colon \mathsf{A} \to \mathsf{A}$ *an additive endo-functor of* A.

Then there exists an additive functor $j\colon (\mathsf{A}, e) \to (\mathsf{A}^\sharp, t)$ *into an additive category* $\mathsf{A}^\sharp$ *together with an (additive)* auto-equivalence $t\colon \mathsf{A}^\sharp \to \mathsf{A}^\sharp$ *such that:*

(i) j *transforms* e *into* t^{-1}*, i.e.:* $t^{-1}j$ *is isomorphic to* $j \cdot e$,

(ii) *any additive functor* $F\colon (\mathsf{A}, e) \to (\mathsf{B}, s)$ *into an additive category* B *with auto-equivalence* s*, which transforms* e *into* s^{-1}*, factors uniquely over* j *by an additive functor* $F^\sharp\colon (\mathsf{A}^\sharp, t) \to (\mathsf{B}, s)$ *which transforms* t *into* s*, that is* $F = F^\sharp j$ *and* $F^\sharp t$ *is isomorphic to* $sF^\sharp$.

48 *The functor* j *is unique up to "equivalence of categories with auto-equivalences under* (A, e)*".*

In other words—beware of the universe—there is a left adjoint of the forgetful functor from (Ad, aut)*, the "category of additive categories with auto-equivalence" to* (Ad, end)*, the "category of additive categories with endo-functors".*

PROOF. The *objects* of $\mathsf{A}^\sharp$ are given by all pairs (X, n), where X is an object in A, n an integer. The *morphisms* are given by

$$\mathrm{Hom}_{\mathsf{A}^\sharp}((X, n), (Y, m)) = \varinjlim_{i \geq -m, -n} \mathrm{Hom}_{\mathsf{A}}(e^{i+m}X, e^{i+n}Y)\,,$$

the *auto-equivalence* t by $t(X, n) = (X, n+1)$ and the identity on the direct limits representation the morphisms.

The functor j sends an object X from A to $(X, 0)$ and is given on morphisms by the natural map of $\mathrm{Hom}_{\mathsf{A}}(X, Y)$ into the direct limit $\varinjlim_{i \geq 0} \mathrm{Hom}_{\mathsf{A}}(e^iX, e^iY)$.

If $F\colon (\mathsf{A}, e) \to (\mathsf{B}, s)$ is now a functor as in (ii), set

$$F^\sharp(X, n) = s^n F(X)\,.$$

It is left to the reader to convince themself that these data solve the universal mapping problem claimed—cf. also [**BBD82**, 1.1.5.]. □

Coming back to the category of S-modules, it was stated in (2.2.2) that the composition of the natural embedding of $\mathrm{mod}(S)$ into $\mathrm{D^b}(S)$ with the projection from $\mathrm{D^b}(S)$ into $\underline{\mathrm{D}}^{\mathrm{b}}(S)$ factors over the functor ι_S, which is defined on $\underline{\mathrm{mod}}(S)$ and transforms the loop-space functor into the inverse T^{-1} of the translation functor on $\underline{\mathrm{D}}^{\mathrm{b}}(S)$. Hence $\iota_S\colon \underline{\mathrm{mod}}(S) \to \underline{\mathrm{D}}^{\mathrm{b}}(S)$ factors over the category $(\underline{\mathrm{mod}}(S)^\sharp, \Sigma_S)$, constructed from $\underline{\mathrm{mod}}(S)$ by "inverting Ω_S" in the sense above—so that Σ_S represents the suspension functor. The induced functor $\iota_S^\sharp$ transforms then the suspension functor into the translation functor and Proposition 6.5.1 is equivalent to:

THEOREM 6.5.3. $\iota_S^\sharp\colon (\underline{\mathrm{mod}}(S)^\sharp, \Sigma_S) \to (\underline{\mathrm{D}}^{\mathrm{b}}(S), \mathrm{T})$ *is an equivalence of graded categories,* [**Ver77**, 1.1.0.].[5] □

In particular, $\underline{\mathrm{mod}}(S)^\sharp$ inherits a triangulated structure via this equivalence and the reader is asked to describe the distinguished triangles using—and extending—(4.7).

[5]Compare this with ((4.8.1)) outside of the strongly Gorenstein case.

CHAPTER 7

Multiplicative structure, duality and support

So far, we have dealt with the definition and properties of Tate cohomology for 49
a fixed ring S which is strongly Gorenstein.

The main aspect here will be the functorial behavior of the cohomology theory, culminating with a general duality theorem and its applications, which encompass the classical such theorem for Tate cohomology of integral group rings as well as the analogue of Grothendieck–Serre duality for finite morphisms (of commutative rings).

But we start with a short discussion of

7.1 Yoneda or extension products. The very definition of Tate cohomology as groups of morphisms in the "graded" category $\underline{\mathrm{D}}^{\mathrm{b}}(S)$ implies that the composition of such morphisms yields bilinear pairings (the Yoneda products) for any objects X, Y and Z in $\underline{\mathrm{D}}^{\mathrm{b}}(S)$ and all integers i and j:

$$\underline{\mathrm{Ext}}^i_S(Z,Y)\times\underline{\mathrm{Ext}}^j_S(X,Z)\longrightarrow\underline{\mathrm{Ext}}^{i+j}_S(X,Y)\,, \tag{7.1.1}$$

functorial in its arguments and associative in the usual sense.

As the natural transformation $c\colon \mathrm{Ext}_S(-,-)\to\underline{\mathrm{Ext}}_S(-,-)$ is induced by the projection functor from $\mathrm{D}^{\mathrm{b}}(S)$ onto $\underline{\mathrm{D}}^{\mathrm{b}}(S)$ by (6.2.5) 3, it is necessarily compatible with these products.

Considering the behavior of the products with respect to duality, one has as usual—see [**AGV72**, XVII.1.1]—that the following diagram is commutative, up to the sign $(-1)^{ij}$, (where $(-)^*$ represents $\mathrm{RHom}_S(-,S)$ as in (6.2.3), the horizontal arrows are given by the respective products, the vertical isomorphisms by "transposition" as in (6.2.5), followed on the left hand side by the exchange of factors):

$$\begin{array}{ccc}
\underline{\mathrm{Ext}}^i_S(Z,Y)\times\underline{\mathrm{Ext}}^j_S(X,Z) & \longrightarrow & \underline{\mathrm{Ext}}^{i+j}_S(X,Y)\\
\cong\updownarrow & (-1)^{ij} & \updownarrow\cong\\
\underline{\mathrm{Ext}}^j_{S^{\mathrm{op}}}(Z^*,X^*)\times\underline{\mathrm{Ext}}^i_{S^{\mathrm{op}}}(Y^*,Z^*) & \longrightarrow & \underline{\mathrm{Ext}}^{i+j}_{S^{\mathrm{op}}}(Y^*,X^*)\,.
\end{array} \tag{7.1.2}$$

In other words, for any two given classes f in $\underline{\mathrm{Ext}}^i_S(Z,Y)$, g in $\underline{\mathrm{Ext}}^j_S(X,Z)$, one has 50

$$(f\circ g)^*=(-1)^{ij}(g^*\circ f^*)\quad\text{in}\quad\underline{\mathrm{Ext}}^{i+j}_{S^{\mathrm{op}}}(Y^*,X^*)\,.$$

Recall furthermore, [**Bou80**, p128], that on the (hyper-)Tor-groups $\mathrm{Tor}^S(Y,X^*)$—cf. (6.2.3) and 6.2.4—there are natural *left* actions of both $\mathrm{Ext}_S(Y,Z)$ and $\mathrm{Ext}_{S^{\mathrm{op}}}(X^*,W^*)$ for all objects Z, W in $\mathrm{D}^{\mathrm{b}}(S)$.

These actions *anti-commute*, or, equivalently, if we define a *right* action of $\mathrm{Ext}_S(W,X)$ on $\mathrm{Tor}^S(Y,X^*)$:

$$\mathrm{Tor}_i^S(Y,X^*) \times \mathrm{Ext}_S^j(W,X) \longrightarrow \mathrm{Tor}_{i-j}^S(Y,W^*)$$
$$\text{by} \qquad x \circ f := (-1)^{ij}(f^* \circ x)\,,$$

for all integers i and j, this newly defined action of $\mathrm{Ext}_S(W,X)$ on $\mathrm{Tor}^S(Y,X^*)$ becomes compatible with the original left action of $\mathrm{Ext}_S(Y,Z)$.

Putting all these different actions together, we get:

PROPOSITION 7.1.3. *Let W, X, Y, Z be objects in $\mathrm{D^b}(S)$. Then the morphisms in the exact sequence* (6.2.5) (3) *for X and Y are*

- *right-linear with respect to* $\mathrm{Ext}_S(W,X)$ *and*
- *left-linear with respect to* $\mathrm{Ext}_S(Y,Z)$.

More precisely, given classes f in $\mathrm{Ext}_S^i(Y,Z)$ and g in $\mathrm{Ext}_S^j(W,X)$, right multiplication with g (resp. $c^j(g)$) and left multiplication with f (resp. $c^i(f)$) yields a homomorphism of exact sequences

$$\begin{array}{ccccccc}
\cdots \longrightarrow & \underline{\mathrm{Ext}}_S^{k-1}(X,Y) & \longrightarrow & \mathrm{Tor}_{-k}^S(Y,X^*) & \longrightarrow & \mathrm{Ext}_S^k(X,Y) & \xrightarrow{c^k} \cdots \\
& \downarrow{\scriptstyle c^i(f)\circ\,|\,\circ c^j(g)} & & \downarrow{\scriptstyle f\circ\,|\,\circ g} & & \downarrow{\scriptstyle f\circ\,|\,\circ g} & \\
\cdots \longrightarrow & \underline{\mathrm{Ext}}_S^{i+j+k-1}(W,Z) & \longrightarrow & \mathrm{Tor}_{-i-j-k}^S(Z,W^*) & \longrightarrow & \mathrm{Ext}_S^{i+j+k}(W,Z) & \xrightarrow{c^{i+j+k}} \cdots\,.
\end{array}$$

Taking into account the already observed degeneracy of these exact sequences, (6.3.4), one can further analyse the structure of the Norm map $\mathbf{N}(X,Y)$, (6.2.5) 1.

51 EXAMPLE 7.1.4. Let x be an element of $\mathrm{Ext}_S^k(X,Y)$. If there is a pair (Z,f) with Z an object in $\mathrm{D^b}(S)$ and f an element in $\mathrm{Ext}_S^i(Y,Z)$ such that

$$k+i > \overline{m}(X,Z)\,, \quad \text{(with notations as in (6.3.3)), and}$$
$$f \circ x \neq 0 \quad \text{in} \quad \mathrm{Ext}_S^{k+i}(X,Z)\,,$$

then necessarily:

$$c^k(x) \neq 0 \quad \text{in} \quad \underline{\mathrm{Ext}}_S^k(X,Y)\,,$$

that is: x is not in the image of $\mathrm{H}(\mathbf{N}(X,Y))$, as

$$c^i(f) \circ c^k(x) = c^{i+k}(f \circ x)\,,$$

and c^{i+k} is an isomorphism by the first assumption and (6.3.4) (i).

Hence, in a sense which is left to be made precise by the reader, the (cohomological) Norm map $\mathrm{H}(\mathbf{N}(X,Y))\colon \mathrm{Tor}^S(Y,X^*) \to \mathrm{Ext}_S(X,Y)$ takes its values only in the subgroup of those classes of $\mathrm{Ext}_S(X,Y)$ which get annihilated by the Yoneda product with any class of sufficiently high degree. (If one considers in analogy to the geometric situation the projection of $\mathrm{D^b}(S)$ onto $\underline{\mathrm{D}}^{\mathrm{b}}(S)$ as a "localization functor"—a point of view which is supported by (6.5.3)—one may say that $\mathrm{H}(\mathbf{N}(X,Y))$ takes its values in the "zeroth local cohomology" of sections with support in $\mathrm{D^b_{perf}}(S)$, off which one localizes.)

It seems rather suggestive to believe that these classes constitute in fact the image of the cohomological Norm map.

Similarly, its kernel contains all elements which are "eventually decomposable" in the appropriate (obvious) sense.

To conclude this subsection, specialise (7.1.1) to the case where $X = Y = Z$. This yields for any object X in $\mathrm{D^b}(S)$ a "stabilized Yoneda-Ext-algebra" $\underline{\mathrm{Ext}}_S(X,X)$, which is a $\mathbb{Z}$-graded ring, canonically isomorphic to the *graded opposite* ring of $\underline{\mathrm{Ext}}_{S^{\mathrm{op}}}(X^*,X^*)$.

If X is in fact (isomorphic to) a single S-module M, this algebra comes further equipped with an ascending filtration by subgroups—see (6.5.1)—:

(7.1.5) $$F_i^r(M) = \mathrm{Im}(\mathrm{Ext}_S^r(\Omega_S^i M, \Omega_S^i M) \longrightarrow \underline{\mathrm{Ext}}_S^r(M,M))\,,$$

and this filtration is "good" in the sense that

- $F_i^r(M)\cdot F_j^s(M)$ is contained in $F_{i+j}^{r+s}(M)$ for all i,j,r,s and
- one has equality for sufficiently large i,j.[1]

52

7.2 Operations of a perfect complex. Assume S and T are rings which are both strongly Gorenstein. Consider a complex L of S-T-bimodules, *perfect* as a complex of *left* S-modules, and as a complex *right* T-modules.

Then the derived tensor product $-\otimes_S^{\mathsf{L}} L$ provides an exact functor from $\mathrm{D^b}(S)$ to $\mathrm{D^b}(T)$ which carries perfect complexes into such. Hence it passes trivially to the respective quotients and induces an exact functor between the stable categories

(7.2.1) $$-\underline{\otimes}_S^{\mathsf{L}} L\colon \underline{\mathrm{D}}^{\mathrm{b}}(S) \longrightarrow \underline{\mathrm{D}}^{\mathrm{b}}(T)\,.$$

It admits a right adjoint iff $\mathrm{RHom}_T(L,T)$ is still perfect as a complex of *right S-modules.* In this case, the adjoint is the exact functor

(7.2.2) $$\underline{\mathrm{RHom}}_T(L,-)\colon \underline{\mathrm{D}}^{\mathrm{b}}(T) \longrightarrow \underline{\mathrm{D}}^{\mathrm{b}}(S)$$

which is induced from $\mathrm{RHom}_T(L,-)$ by passing to the quotient-categories.

These functors are then most easily described on $\underline{\mathrm{APC}}(-)$: If A is an acyclic complex of finitely generated projective (right) S-modules—hence an object of $\underline{\mathrm{APC}}(S)$—the total complex associated to $A\otimes_S L$, considered as a complex of T-modules, is necessarily an object of $\underline{\mathrm{APC}}(T)$ and represents naturally $(\underline{\sigma}_{\leqslant} A)\underline{\otimes}_S^{\mathsf{L}} L$.[2]

In the same way, one has a natural identification—cf. (6.2.1) (ii)—

$$\underline{\mathrm{RHom}}_T(L, \underline{\sigma}_{\leqslant} B) = \underline{\sigma}_{\leqslant}(B \otimes_T \mathrm{RHom}_T(L,T))$$

for any complex B in $\underline{\mathrm{APC}}(T)$.

Rewriting this in terms of complete resolutions, it says that there are functorial isomorphisms[3]

$$\mathbf{CR}_S(X)\otimes_S L = \mathbf{CR}_T(X\otimes_S^{\mathsf{L}} L)\,,$$
$$\mathbf{CR}_S(\mathrm{RHom}_T(L,Y)) = \mathrm{Hom}_T(L, \mathbf{CR}_T(Y))$$

for X in $\mathrm{D^b}(S)$, Y in $\mathrm{D^b}(T)$.

The reader should observe that the notation "$\underline{\mathrm{RHom}}_T(L,-)$" is somewhat mis- 53
leading: There are no cohomology groups naturally attached to $\underline{\mathrm{RHom}}_T(L,-)$, in particular $\underline{\mathrm{Ext}}_T(L,-)$, (which groups vanish anyway as L is perfect over T), does *not* represent its cohomology. (As an example, remark that for $S = T = L$, the functor $\underline{\mathrm{RHom}}_S(S,-)$ is naturally isomorphic to the identity functor on $\underline{\mathrm{D}}^{\mathrm{b}}(S)$.)

---

[1] As stated this would imply that the positive part of $\underline{\mathrm{Ext}}$ is noetherian, which it need not be. What is true is that for a fixed $r\geq 0$, $F_i^r(M) = \underline{\mathrm{Ext}}_S^r(M,M)$ for sufficiently large i.

[2] In other words $(\underline{\sigma}_{\leqslant} A)\underline{\otimes}_S^{\mathsf{L}} L \simeq \underline{\sigma}_{\leqslant}(A\otimes_S L)$ in $\underline{\mathrm{D}}^{\mathrm{b}}(S)$.

[3] In these isomorphisms it is necessary that L be represented as a complex of projective modules. We also caution that later $\underline{\mathrm{RHom}}$ is used with a different meaning.

Instead, consider the cohomology groups

$$\mathbf{E} = \underline{\mathrm{Ext}}_T(X \otimes^{\mathsf{L}}_S L, Y) = \underline{\mathrm{Ext}}_S(X, \mathrm{RHom}_T(L, Y)).$$

They are obtained as the common abutment of the two spectral sequences

$$'E_2^{i,j} = \underline{\mathrm{Ext}}^i_T(\mathrm{Tor}^S_j(X, L), Y) \implies \mathbf{E}^{i+j}$$

and

$$''E_2^{i,j} = \underline{\mathrm{Ext}}^i_T(X, \mathrm{Ext}^j_T(L, Y)) \implies \mathbf{E}^{i+j}, \tag{7.2.3}$$

which depend functorially on X in $\mathrm{D^b}(S)$ and Y in $\mathrm{D^b}(T)$.

Furthermore, with respect to the natural transformations c, these spectral sequences "stabilize" (and extend) the corresponding ones for the "ordinary" Ext's.

7.3 Change of rings. Let $f\colon T \to S$ in the following be a ring homomorphism between two rings which are both strongly Gorenstein.

(7.3.1) Then certainly $f^* := - \otimes^{\mathsf{L}}_T S$ transforms perfect complexes over T into such over S. But, almost by definition, it induces a functor from $\mathrm{D^b}(T)$ into $\mathrm{D^b}(S)$—and not just $\mathrm{D}^-(S)$—only if S considered as a left T-module (by restriction of scalars along f) is of *finite flat dimension.* In this case, f^* passes trivially to the quotients and the induced exact functor will be denoted $\underline{f}^*\colon \underline{\mathrm{D}}^{\mathrm{b}}(T) \to \underline{\mathrm{D}}^{\mathrm{b}}(S)$.

(7.3.2) The functor f_*, which "restricts the scalars", will transform $\mathrm{D^b}(S)$ into $\mathrm{D^b}(T)$ iff f_*S, the underlying (right) T-module of S, is *finitely generated* over T. As S generates $\mathrm{D^b_{perf}}(S)$, (1.2.1), f_* will preserve perfectness iff furthermore f_*S is of *finite projective dimension* over T as a *right* module.

54 If in particular S is *perfect,* (that is, finitely generated and of finite projective dimension), on both sides over T—with respect to f—there is a naturally induced pair of adjoint functors $(\underline{f}^*, \underline{f}_*)$ between $\underline{\mathrm{D}}^{\mathrm{b}}(T)$ and $\underline{\mathrm{D}}^{\mathrm{b}}(S)$. This is of course also a special case of (7.2).

(7.3.3) The right adjoint $f^!$ of f_*, obtained as the derived functor of $\mathrm{Hom}_T(f_*S, -)$, transforms modules of finite injective dimension into (complexes isomorphic to) bounded complexes of injective modules, but does not necessarily preserve finite generation, so that in general $\mathrm{D^b}(T)$ is not carried into $\mathrm{D^b}(S)$.

But, if $f^!T = \mathrm{RHom}_T(f_*S, T)$ is *perfect*—as a complex of right modules over S—one may apply (7.2) to S considered as a complex of S-T-bimodules, which yields that $f^!$ induces an exact functor

$$\underline{f}^!\colon \underline{\mathrm{D}}^{\mathrm{b}}(T) \longrightarrow \underline{\mathrm{D}}^{\mathrm{b}}(S),$$

so that the pair $(\underline{f}_*, \underline{f}^!)$ is adjoint again.

(7.3.4) Summarising, in the most favorable case where

- f_*S is perfect on both sides over T and
- $\mathrm{RHom}_T(f_*S, T)$ is perfect as a complex of right S-modules,

there is a trivially induced triple of adjoint functors $(\underline{f}^*, \underline{f}_*, \underline{f}^!)$ between $\underline{\mathrm{D}}^{\mathrm{b}}(S)$ and $\underline{\mathrm{D}}^{\mathrm{b}}(T)$.

It is left to the reader to specialise (7.2) accordingly to obtain the corresponding "stabilized" change-of-rings spectral sequences.

Even under the foregoing strong hypotheses on f, there is still a shortcoming: Without further restrictions—like S and T both being commutative—we ignore the behavior of the deduced triple of adjoint functors with respect to the dualities induced by dualizing modules for S or T, (4.6.1). Hence it seems rather doubtful that the class of "bi-perfect" ring homomorphisms constitutes already the appropriate class of morphisms to turn (strongly) Gorenstein rings into a suitable category.

Furthermore, the following suggests that the above class of ring homomorphisms might be too small:

55

7.4 An extension of Tate cohomology. To put the analogue of "Grothendieck–Serre duality" for Tate cohomology into its proper general context, it will be convenient first to extend the definition of the stabilized $\underline{\text{Ext}}$-groups, allowing arbitrary modules as covariant argument.

This also constitutes the "right" generalisation of the classical case, in which the contravariant argument is the fixed augmentation module, whereas the covariant argument remains unrestricted.

The main reason to treat Tate cohomology here up to now as an essentially symmetric theory was provided by its behavior with respect to the "trivial duality" of (6.2.5) 2, based on (4.6.1), which seemingly places contravariant and covariant arguments on the same footing. But, examining (6.2.5) 3 or (6.4) a little bit closer, it becomes evident that in fact the contravariant argument is special: Fixing, say, a maximal Cohen–Macaulay S-module M and considering (6.4.1) (i)–(ii) as a definition, one may interpret the family of functors $\underline{\text{Ext}}^i_S(M, -)$ to yield a (covariant) cohomological δ-functor on all S-modules—and not just on the finitely generated ones.

To formulate a more precise statement, we will use—and extend—the description given in Remark (b) following (6.2.5). Following [**Ver77**, II.1.1], let us recall that for an arbitrary ring S and $* = -$ or $* = \mathrm{b}$,

$$\mathrm{D}^*(\mathrm{Mod}(S)) = \mathrm{K}^{-,*}(\mathrm{Mod}(S))/\,\mathrm{K}^{-,\varnothing}(\mathrm{Mod}(S))$$

denotes the derived category which is obtained from $\mathrm{K}^{-,*}(\mathrm{Mod}(S))$, the homotopy category of all bounded above complexes of arbitrary right S-modules (with bounded cohomology), by factoring out its thick subcategory $\mathrm{K}^{-,\varnothing}(\mathrm{Mod}(S))$ of all acyclic complexes. It is over such a category that the covariant argument will range.

Assuming again that S is a ring which is strongly Gorenstein, we have

Definition and Proposition 7.4.1. *For any complexes Y in $\mathrm{D}^-(\mathrm{Mod}(S))$ and X in $\mathrm{D}^{\mathrm{b}}(S)$ set*

$$\underline{\mathrm{RHom}}_S(X, Y) = \mathrm{Hom}_S(\mathbf{CR}(X), Y)\,.$$

(i) *This defines a bi-functor into the (full) derived category of abelian groups,*

$$\underline{\mathrm{RHom}}_S(-,-)\colon \underline{\mathrm{D}}^{\mathrm{b}}(S)^{\mathrm{op}} \times \mathrm{D}^-(\mathrm{Mod}(S)) \longrightarrow \mathrm{D}(\mathsf{Ab})\,,$$

which is exact in both arguments. 56

(ii) *The corresponding cohomology groups are still denoted*

$$\underline{\mathrm{Ext}}^i_S(X, Y) := \mathrm{H}^i\,\mathrm{Hom}_S(\mathbf{CR}(X), Y)$$

and they coincide naturally with the Tate cohomology groups as defined in (6.1.1) if Y has in fact bounded and finitely generated cohomology only.

(iii) *For* Y *in* $\mathrm{D^b}(\mathrm{Mod}(S))$ *there is a convergent spectral sequence*

$$E_2^{i,j} = \underline{\mathrm{Ext}}_S^i(\mathbf{M}(X), \mathrm{H}^j(Y)) \Longrightarrow \underline{\mathrm{Ext}}_S^{i+j}(X,Y)\,,$$

whose E_2*-terms can be calculated using* (6.4.1).

PROOF. As stated, the functor $\underline{\mathrm{RHom}}(-,-)$ is a priori defined on

$$\mathrm{K}^{-,-}(\mathrm{Mod}(S)) \times \underline{\mathrm{APC}}(S)^{\mathrm{op}}$$

and is obviously exact in each argument.

Given that $\mathbf{CR}(-)$ induces an equivalence between $\underline{\mathrm{D}}^{\mathrm{b}}(S)$ and $\underline{\mathrm{APC}}(S)$ by (5.6) and (4.4.1) (1), it remains only to be seen that as soon as Y is an acyclic, bounded above complex of (arbitrary right) S-modules, the complex of groups $\mathrm{Hom}_S(\mathbf{CR}(X), Y)$ is also acyclic.

But this is well-known (and easily established) for *any*, not necessarily acyclic, complex of projectives in the contravariant argument, hence holds a fortiori in the special case under consideration.

That this definition indeed extends (6.1.1), follows from Remark (b) of (6.2.5).

The last assertion—which we will not use—is left as an exercise. □

REMARK. (a) This definition is certainly *not* symmetric in its arguments in the sense that even if Y is in $\mathrm{D^b}(\mathrm{Mod}(S))$, one cannot derive the covariant argument first by replacing it with an injective resolution:

If I is a bounded below complex of injective modules, X an arbitrary acyclic complex (for example a complete resolution), then the complex $\mathrm{Hom}_S(X, I)$ is acyclic! In other terms, identifying $\mathrm{D^b}(\mathrm{Mod}(S))$ with

$$\mathrm{K}^{+,\mathrm{b}}(\mathrm{Mod}(S))/\,\mathrm{K}^{+,\varnothing}(\mathrm{Mod}(S))\,,$$

the functor $\underline{\mathrm{RHom}}_S(-,-)$ is not obtainable by passage to the quotient from a functor defined on $\mathrm{K}^{+,\mathrm{b}}(\mathrm{Mod}(S))$ in its covariant argument.[4]

(b) One may still extend the definition of the Yoneda products to this situation: The original stabilized $\underline{\mathrm{Ext}}$s act from the right, the ordinary Exts (on $\mathrm{D}^-(\mathrm{Mod}(S))$) act from the left. This is obvious.

57 A first reason to extend the definition of the Tate cohomology groups is that they behave (slightly) better with respect to ring homomorphisms:

PROPOSITION 7.4.2. *Let* $f\colon T \to S$ *be a homomorphism of rings which are both strongly Gorenstein.*

(i) *If* S *is of finite flat dimension as a* left T*-module, there is a natural isomorphism of (bi-)functors on* $\underline{\mathrm{D}}^{\mathrm{b}}(T)^{\mathrm{op}} \times \mathrm{D}^-(\mathrm{Mod}(S))$

$$\underline{\mathrm{RHom}}_S(\underline{f}^*-,-) = \mathrm{RHom}_T(-, f_*-)\,.$$

(ii) *If* S *is finitely generated and of finite projective dimension as a right* T*-module, there is a natural isomorphism of (bi-)functors on* $\underline{\mathrm{D}}^{\mathrm{b}}(S)^{\mathrm{op}} \times \mathrm{D}^-(\mathrm{Mod}(T))$

$$\underline{\mathrm{RHom}}_T(\underline{f}_*-,-) = \mathrm{RHom}_S(-, f^!-)\,.$$

The proof follows immediately from (7.3). The hypotheses just guarantee that the underlined functors are defined.

---

[4] All of this is true but nonetheless it is possible to extend the Tate cohomology groups to include large modules in the contravariant argument. The large stable derived category (which has $\underline{\mathrm{D}}^{\mathrm{b}}(S)$ as its subcategory of compact objects) was constructed in [**88**].

7.5 The Duality Theorem. Using the now extended definition of the Tate cohomology groups, we can formulate the rather broad version of a

DUALITY THEOREM 7.5.1. Assume the following conditions to hold:

- S is a ring which is strongly Gorenstein, T is an arbitrary ring,
- M is a complex in $\mathrm{D^b}(S^{\mathrm{op}})$,
- N is a bounded above complex of S-T-bimodules with only finitely many non-zero cohomology modules. (Hence it represents an object in $\mathrm{D^b}(\mathrm{Mod}(S^{\mathrm{op}}\otimes T))$.)
- W is a complex of right T-modules, quasi-isomorphic to a *finite* complex of right *injective* T-modules.

Then there are two spectral sequences converging to the same limit $\mathbf{E}$:

$$'E_2^{i,j} = \mathrm{Ext}^i_T(\underline{\mathrm{Ext}}^{-j}_{S^{\mathrm{op}}}(M, N), W) \Longrightarrow \mathbf{E}^{i+j}$$

and

$$''E_2^{i,j} = \underline{\mathrm{Ext}}^{i-1}_S(M^*, \mathrm{Ext}^j_T(N, W)) \Longrightarrow \mathbf{E}^{i+j}\,.$$

(Here again M^* is short for $\mathrm{RHom}_{S^{\mathrm{op}}}(M, S)$, cf. (6.2.3).)

PROOF. Let $W \to I$ be an injective resolution of W over T. By assumption, I can—and will be—chosen to be a bounded complex. The common abutment of the two spectral sequences will be given by the cohomology of the total complex associated to the double complex

$$C = \mathrm{Hom}_T(\mathrm{Hom}_{S^{\mathrm{op}}}(\mathbf{CR}(M), N), I)\,.$$

More precisely, its terms are given by 60[5]

$$C^{i,j} = \mathrm{Hom}^i_T(\mathrm{Hom}^{-j}_{S^{\mathrm{op}}}(\mathbf{CR}(M), N), I) = \mathrm{Hom}_T(\prod_k \mathrm{Hom}_{S^{\mathrm{op}}}(\mathbf{CR}^k(M), N^{k-j}), I^i)\,.$$

Now, as I is supposed to be finite, both spectral sequences associated to this double complex are biregular, [**Gro66**, 0.11.3.3.(ii)], and the $'E_2$-terms of the spectral sequence associated to the first filtration of this double complex are the groups $'E_2^{i,j}$ of the theorem as I is assumed to be an injective resolution of W.

Here we did not need that N has bounded cohomology. This condition only intervenes in dealing with the second spectral sequence.

For this, remark first that the cohomology of (the total complex associated to) C does not change if one replaces N by a quasi-isomorphic complex in

$$\mathrm{K}^{-,-}(\mathrm{Mod}(S^{\mathrm{op}} \otimes T))\,:$$

That $\mathrm{Hom}_S(\mathbf{CR}(M), -)$ preserves such quasi-isomorphisms was explained above, (7.4.1), and for $\mathrm{Hom}_T(-, I)$ it is obvious. Hence we may and will assume that N is indeed a bounded complex of S-T-bimodules.

Then, in the foregoing explicit description of $C^{i,j}$ the product may be replaced by a (finite) direct sum so that the natural morphism of double complexes

$$N \otimes_{S^{\mathrm{op}}} \mathrm{Hom}_{S^{\mathrm{op}}}(\mathbf{CR}(M), S) \longrightarrow \mathrm{Hom}_{S^{\mathrm{op}}}(\mathbf{CR}(M), N)$$

is in fact an isomorphism, compatible with the right T-module structures on either side.

[5]In the unpublished version of these notes (available from the University of Toronto T-space), the pages are out of order here. In those notes there is a page missing between p57 and p58, which shows up later on p60.

Composing this with the "interchange-isomorphism" of double complexes—[**Bou80**, 71]—:

$$\mathrm{Hom}_{S^{\mathrm{op}}}(\mathbf{CR}(M), S) \otimes_S N \longrightarrow N \otimes_{S^{\mathrm{op}}} \mathrm{Hom}_{S^{\mathrm{op}}}(\mathbf{CR}(M), S)\,,$$

and passing to the associated total complexes, one identifies the double complex C with the double complex

$$'C = \mathrm{Hom}_T(\mathbf{CR}(M)^* \otimes_S N, I)\,.$$

Here, as before, $(-)^*$ denotes the dual—this time with respect to S^{op}.

Now apply finally to $'C$ term by term the adjunction isomorphism which induces an isomorphism of the total complexes associated:

$$\mathrm{Hom}_T(\mathbf{CR}(M)^* \otimes_S N, I) = \mathrm{Hom}_S(\mathbf{CR}(M)^*, \mathrm{Hom}_T(N, I))$$

and observe that $\mathbf{CR}(M)^*$ is canonically isomorphic to $\mathbf{CR}(M^*)[1]$ by (4.6.1).

Putting all this together, it follows that the total complexes associated to C and

$$''C = \mathrm{Hom}_S(\mathbf{CR}(M^*)[1], \mathrm{Hom}_T(N, I))$$

are naturally isomorphic.

58 Remark that the terms of the double complex $''C$ are given by

$$''C^{i,j} = \mathrm{Hom}_S(\mathbf{CR}^{-i+1}(M^*), \prod_k \mathrm{Hom}_T(N^k, I^{k+j}))\,.$$

As $\mathrm{Hom}_T(N, I)$ is bounded (above and below) by the assumptions on N and I, the natural filtration on $\mathbf{CR}(M^*)[1]$ induces a regular filtration on $''C$, and the E_2-terms of the associated convergent spectral sequence are precisely the groups $''E_2^{i,j}$ of the theorem, as follows from (7.4.1) (ii), taking into account that the components of $\mathbf{CR}(M^*)$ are finitely generated projective S-modules.

This finishes the proof of (7.5.1). □

Let us immediately record the simplest case in which both spectral sequences of (7.5.1) are degenerate, (and which generalises for example the corresponding well-known statement for integral group rings—see [**CE99**, XII.6.5.]):

COROLLARY 7.5.2. *Maintaining the assumptions and notations of the foregoing theorem, assume furthermore that $W = I$ is actually already an injective T-module. Then each of the two spectral sequences in* (7.5.1) *collapses, the edge homomorphisms $''E_2^{j,0} \to \mathbf{E}^j$ and $\mathbf{E}^j \to {}'E_2^{0,j}$ are defined and are isomorphisms of abelian groups for any integer j.* □

More suggestively, let us denote by $\mathbf{D}(-)$ either contravariant functor on $\mathrm{D}(\mathrm{Mod}(T))$ or $\mathrm{D}(\mathrm{Mod}(S^{\mathrm{op}} \otimes T))$ induced from $\mathrm{Hom}_T(-, I)$. Then the composition of the aforementioned edge homomorphisms yields an isomorphism of abelian groups for any integer j:

$$\underline{\mathrm{Ext}}_S^{j-1}(M^*, \mathbf{D}(N)) \xrightarrow{\cong} \mathbf{D}(\underline{\mathrm{Ext}}_{S^{\mathrm{op}}}^{-j}(M, N))\,, \tag{7.5.3}$$

and these isomorphisms are natural in M and N.

Before giving more sophisticated applications of the Duality Theorem, we want to analyse a little bit closer its proof:

59 All isomorphisms occurring above in the identification of the total complexes associated to C, respectively $''C$, are completely natural—except for the "interchange isomorphism" which presupposes the choice of a "commutation factor" and

introduces signs. These signs appear if one studies the behavior of this duality with respect to Yoneda products.

In the following discussion, assume N and W in (7.5.1) to be fixed and consider the variation of the spectral sequences in M.

If L is another object in $\mathrm{D^b}(S^{\mathrm{op}})$, the Yoneda product from the *right* with $\underline{\mathrm{Ext}}^k_S(L^*, M^*)$ maps

$$
\begin{aligned}
{}''E_2^{i,j}(M) &= \underline{\mathrm{Ext}}_S^{i-1}(M^*, \mathrm{Ext}^j_T(N,W)) \quad \text{into} \\
{}''E_2^{i+k,j}(L) &= \underline{\mathrm{Ext}}_S^{i+k-1}(L^*, \mathrm{Ext}^j_T(N,W)),
\end{aligned}
$$

as well as $\mathbf{E}^i(M)$ into $\mathbf{E}^{i+k}(L)$—consider these groups as the cohomology of the total complexes associated to $''C(M)$ or $''C(L)$ respectively.

This action is natural in the sense that it commutes with the respective differentials of the (doubly primed) spectral sequences and is compatible with the induced filtrations on the limit terms. In other words, any given element of $\underline{\mathrm{Ext}}^k_S(L^*, M^*)$ defines in a natural way a morphism of spectral sequences of bi-degree $(k, 0)$.

Dually, $\underline{\mathrm{Ext}}^k_{S^{\mathrm{op}}}(M, L)$ acts from the right on $\underline{\mathrm{Ext}}_{S^{\mathrm{op}}}(L, N)$, and applying $\mathrm{Ext}^i_T(-, W)$ to it, it furnishes a natural *left* action of this group which maps

$$
\begin{aligned}
{}'E_2^{i,j}(M) &= \mathrm{Ext}^i_T(\underline{\mathrm{Ext}}^{-j}_{S^{\mathrm{op}}}(M,N), W) \quad \text{into} \\
{}'E_2^{i,j+k}(L) &= \mathrm{Ext}^i_T(\underline{\mathrm{Ext}}^{-j-k}_{S^{\mathrm{op}}}(L,N), W).
\end{aligned}
$$

Chasing through the identifications in the proof above and taking into account (7.1.2), the connection between these two actions can be expressed as follows:

COROLLARY 7.5.4. [6] *With the foregoing notations, if α is a class in $\mathrm{Ext}^k_{S^{\mathrm{op}}}(M, L)$, and α^* is its transpose in $\mathrm{Ext}^k_S(L^*, M^*)$, there is the following diagram of morphisms of spectral sequences, commutative in an obvious sense:*

$$
\begin{array}{ccccc}
\mathrm{Ext}^i_T(\underline{\mathrm{Ext}}^{-j}_{S^{\mathrm{op}}}(M,N), W) & \Longrightarrow & \mathbf{E}^{i+j}(M) & \Longleftarrow & \underline{\mathrm{Ext}}_S^{i-1}(M^*, \mathrm{Ext}^j_T(N,W)) \\
\Big\downarrow {\scriptstyle (-1)^{jk}\,\mathrm{Ext}^i_T(-\circ\alpha, W)} & & \Big\downarrow {\scriptstyle -\circ\alpha^*} & & \Big\downarrow {\scriptstyle -\circ\alpha^*} \\
\mathrm{Ext}^i_T(\underline{\mathrm{Ext}}^{-j-k}_{S^{\mathrm{op}}}(L,N), W) & \Longrightarrow & \mathbf{E}^{i+j+k}(L) & \Longleftarrow & \underline{\mathrm{Ext}}_S^{i+k-1}(L^*, \mathrm{Ext}^j_T(N,W)).
\end{array}
$$

□

61

7.6 To prepare for another application of the Duality Theorem, assume—still with the notations and hypotheses of (7.5.1)—that N is in fact a complex of S-bimodules, whose right T-module structure is given by restricting the scalars on the right with respect to a homomorphism of rings $f\colon T \to S$.

Recall that $\mathrm{Hom}_T(S, -)$, considered as the right adjoint of the forgetful functor f_*, carries injective T-modules into injective modules over S. In particular, for a complex W of T-modules as in (7.5.1), the complex $f^!W$ is quasi-isomorphic to a bounded complex of injective S-modules.

Hence, using the adjunction $(f_*, f^!)$, the spectral sequences in (7.5.1) become

$$
\begin{aligned}
{}'E_2^{i,j} &= \mathrm{Ext}^i_S(\underline{\mathrm{Ext}}^{-j}_{S^{\mathrm{op}}}(M,N), f^!W) \Longrightarrow \mathbf{E}^{i+j} \\
{}''E_2^{i,j} &= \underline{\mathrm{Ext}}_S^{i-1}(M^*, \mathrm{Ext}^j_S(N, f^!W)) \Longrightarrow \mathbf{E}^{i+j},
\end{aligned}
\tag{7.6.1}
$$

[6]In the manuscript this corollary is also labeled 7.5.2.

which, equivalently, are the spectral sequences of (7.5.1) for $T = S$ and $f^!W$ instead of W.

Now we will be interested in the case where W can be chosen to be a dualizing module for T, (and T strongly Gorenstein), such that $f^!W$ is still at least a "dualizing complex" for S—with the purpose that then for N an MCM S-module on the right, the doubly primed spectral sequence above degenerates.

We ignore the general situation in which this can be done, but let us exhibit the following particular one:

DEFINITION 7.6.2. In analogy to the commutative case, let us call a ring homomorphism $f\colon T \to S$ a (central) *duality morphism of finite type* iff

(i) S is strongly Gorenstein and T is a commutative Gorenstein ring (of finite Krull dimension),
(ii) the morphism f turns S into a central T-algebra, such that the underlying T-module f_*S is finitely generated and of finite projective dimension over T, say $d = \operatorname{projdim}_T f_*S$,
(iii) $f_*(-)[-d]$ detects maximal Cohen–Macaulay S-modules in the sense that M in $\operatorname{mod}(S)$ is MCM iff

$$\operatorname{Ext}^i_T(f_*M[-d], T) = \operatorname{Ext}^{i+d}_T(f_*M, T) = 0 \quad \text{for} \quad i \neq 0\,.$$

62 Keeping closely to the usual proof in the commutative case, we will establish, as claimed, that for such duality morphisms

$f^!$ transforms dualizing T-modules into dualizing S-modules

in the following more precise sense:

PROPOSITION 7.6.3. *Assume that $f\colon T \to S$ satisfies* (7.6.2). *Then the following holds:*

(i) *The cohomology of $f^!T$ is concentrated in degree d and*

$$\omega_{S/T} := \mathrm{H}^d(f^!T)$$

is a dualizing module for S.

(ii) *In particular, $\omega_{S/T} \simeq f^!T[d]$ in $\mathrm{D^b}(S)$ and $f^!T$ is a perfect complex of S-modules.*
(iii) *One has $d = \dim T - \operatorname{vdim} S$.*[7]

PROOF. First, take $M = S$ in condition (iii) of (7.6.2), use that S is MCM and that

$$\operatorname{Ext}^i_T(f_*S[-d], T) = \operatorname{Ext}^{i+d}_S(S, f^!T) = \mathrm{H}^{i+d}(f^!T)\,.$$

This shows that the cohomology of $f^!T$ is indeed concentrated in degree d, or, equivalently, that one has $\omega_{S/T} = \mathrm{H}^d(f^!T) \simeq f^!T[d]$ in $\mathrm{D}(\operatorname{Mod}(S))$.

Next, observe that $f^!T$ is in a natural way (represented by) a complex of S-*bimodules* (with bounded cohomology) and that forgetting the hence still existing *left* S-module structure on $\operatorname{RHom}_S(N, f^!T)$ for any given N in $\mathrm{D}(\operatorname{Mod}(S))$ yields a natural isomorphism of complexes of T-modules

$$f_* \operatorname{RHom}_S(N, f^!T) = \operatorname{RHom}_T(f_*N, T)$$

in $\mathrm{D}(\operatorname{Mod}(T))$. (For obvious reasons, we refrain from distinguishing the "left" and "right" forgetful functor.)

[7]This requires an additional hypothesis; see Remark A.7.1.

This rather pedantic remark shows in particular that $\omega_{S/T}$ is in a natural way an S-bimodule and it remains to be seen that it is indeed (projective on both sides and) invertible. For this, we prove:

(a) The (right) S-module $\omega_{S/T}$ is finitely generated as already the T-module

$$f_*\omega_{S/T} = \mathrm{H}^d(f_*f^!T) = \mathrm{H}^d(\mathrm{RHom}_T(f_*S, T))$$

is finitely generated by condition (ii) of (7.6.2).

(b) $\omega_{S/T}$ is of *finite projective dimension* as a (right) S-module: For any S-module M, one has by the above already that **63**

$$\mathrm{Ext}^i_S(M, \omega_{S/T}) = \mathrm{Ext}^i_S(M, f^!T[d]) = \mathrm{Ext}^i_T(f_*M[-d], T)\,.$$

But by condition (iii) of (7.6.2), these groups vanish for $i \neq 0$ if(f) M is MCM and (5.1.1) (ii) applies.

(c) $\omega_{S/T}$ is *MCM over* S: This is essentially the (local) duality theorem for commutative Gorenstein rings. Namely, by the foregoing and (4.6.1), one has

$$\begin{aligned}\mathrm{RHom}_T(f_*\omega_{S/T}[-d], T) &= \mathrm{RHom}_T(f_*f^!T, T)\\ &= \mathrm{RHom}_T(\mathrm{RHom}_T(f_*S, T), T)\\ &= f_*S[0]\,,\end{aligned}$$

whence $\omega_{S/T}$ satisfies the condition in (7.6.2) (iii).

Now apply (5.1.1) (iv) to conclude from (a)–(c) that $\omega_{S/T}$ is finitely generated projective as a right S-module (and by symmetry then also as a left S-module). This already proves (ii).

To get (i) completely, it remains still to verify that $\omega_{S/T}$ is an invertible S-bimodule. But in view of the above and by symmetry again, this reduces to the condition that the natural ring homomorphism

$$S \longrightarrow \mathrm{Hom}_S(\omega_{S/T}, \omega_{S/T})\,,$$

associating to an element of S the left multiplication with it, is an isomorphism of S-bimodules.

It is evidently enough to prove instead that the underlying T-linear map is bijective. This is essentially the chain of identifications in (c) above, read backwards, and taking into account that

$$\begin{aligned}\mathrm{RHom}_T(f_*f^!T, T) &= f_*\,\mathrm{RHom}_S(f^!T, f^!T)\\ &= f_*\,\mathrm{RHom}_S(\omega_{S/T}, \omega_{S/T})[0]\end{aligned}$$

in $\mathrm{D}(\mathrm{Mod}(T))$, $\omega_{S/T} = f^!T[d]$ being projective over S. This finishes the proof of (i).

Finally, we show that (i) $\Rightarrow$ (iii).

As $\omega_{S/T}$ is finitely generated projective over S, one has necessarily

$$\mathrm{injdim}_S\,\omega_{S/T} = \mathrm{vdim}\,S\,.$$

As T is assumed to be Gorenstein, one has $\mathrm{injdim}_T\,T = \dim T$. Then, if I denotes an injective resolution of T as a module over itself, the complex $\mathrm{Hom}_T(S, I)$ represents $f^!T$ in $\mathrm{D^b}(\mathrm{Mod}(T))$

Choosing I of minimal length, this means in view of (i) that $\omega_{S/T} = \mathrm{H}^d(f^!T)$ admits an injective resolution of length at most $\dim T - d$, whence **64**

$$\mathrm{vdim}\,S = \mathrm{injdim}_S\,\omega_{S/T} \leq \dim T - d\,.$$

On the other hand,

$$\mathrm{Ext}^i_T(f_*(-),T) = \mathrm{Ext}^{i-d}_S(-,f^!T[d]) = \mathrm{Ext}^{i-d}_S(-,\omega_{S/T})\,,$$

so that necessarily[8]

$$\mathrm{injdim}_S\,\omega_{S/T} \geq \mathrm{injdim}_T\, T - d = \dim T - d\,,$$

which finishes the proof of the proposition. □

REMARKS. (a) The notation $\omega_{S/T}$for the "relative dualizing module" follows the general (mis-)usage in the commutative case; ω_f would be more accurate.

(b) The proposition implies that $f_*\omega_{S/T}$ is a T-module of finite projective dimension, equal to d. More precisely, choosing a projective resolution $P \to f_*S$ of f_*S over T of (minimal) length d, the object $f_*f^!T[d]$ in $\mathrm{D}(\mathrm{Mod}(T))$ is represented by $\mathrm{Hom}_T(P,T)[d]$, which is then a projective resolution of $f_*\omega_{S/T}$.

This yields the existence of a "projection formula" (see also [**Kle80**]): If N is any complex of T-modules in $\mathrm{D}^-(\mathrm{Mod}(T))$, then

$$f_*f^!(N) = N \otimes^{\mathsf{L}}_T f_*f^!T = (N \otimes^{\mathsf{L}}_T f_*\omega_{S/T})[-d]$$

in $\mathrm{D}^-(\mathrm{Mod}(T))$, and this object is represented by the complex

$$f_*f^!(N) = \mathrm{Hom}_T(P,N) = N \otimes_T \mathrm{Hom}_T(P,T)\,.$$

In particular, if ω_T is any dualizing module for T, then

$$\omega_S := f^!\omega_T[d]$$

represents a dualizing module for S, satisfying

$$f_*\omega_S = \omega_T \otimes^{\mathsf{L}}_T f_*\omega_{S/T}\,.$$

(c) As for examples of such duality morphisms, tighten up (7.6.2) by imposing—in addition to (i) and (ii)—either of the following:

(iii_{f}) S is a (central, finite) flat T-algebra with respect to f, (in which case f_*S is in fact finitely generated projective over T, whence $d = 0$),

or

(iii_{c}) S is also commutative.

It is then well-known that the conditions (i) and (ii) of (7.6.2) together with either[9] (iii_{f}) or (iii_{c}) imply the remaining condition.

65 Having sorted out the above class of "duality morphisms", we need furthermore to recall the following facts from local duality theory for commutative Gorenstein rings.

Let $(R,\mathfrak{m},k)$ be a commutative local Gorenstein ring of Krull dimension r, $\mathfrak{m}$ denoting the unique maximal ideal of R, $k = R/\mathfrak{m}$ the residue class field. If $i\colon k \to E$ denotes an injective envelope of k over R, there exists an injective resolution of R as a module over itself which is of (minimal) length r and exhibits E as its last term. This defines a canonical class γ in $\mathrm{Ext}^r_R(E,R)$ which has the following property:

If A is any R-module of finite length, the (Yoneda-)product (from the left) with γ provides an isomorphism

$$\gamma \circ -\colon\ \mathrm{Hom}_R(A,E) \xrightarrow{\cong} \mathrm{Ext}^r_R(A,R)\,,$$

[8]Not necessarily; see Remark A.7.1.

[9]Each requires an additional hypothesis; see Remark A.7.1.

and for all $i \neq r$ one has $\mathrm{Ext}^i_R(A,R)=0$. In other words, the product with γ yields an isomorphism in $\mathrm{D}(\mathrm{Mod}(S))$ between the single R-module $\mathrm{Hom}_R(A,E)$ and the complex $\mathrm{RHom}_R(A,R)[r]$.

In particular, as $\mathrm{Hom}_R(k,E)$ is isomorphic to k itself, it follows by induction on the length of A that A, $\mathrm{Hom}_R(A,E)$ and $\mathrm{Ext}^r_R(A,R)$ are all R-modules of the same finite length.

Composing the inverse of the isomorphism above with the evaluation-map

$$\mathrm{ev}_A\colon \mathrm{Hom}_R(A,E)\otimes_R A \longrightarrow E\,,$$

one obtains (equivalently) a non-degenerate R-bilinear pairing

$$\langle\ ,\ \rangle_A = \mathrm{ev}_A \circ \big((\gamma\circ -)^{-1}\otimes_R \mathrm{id}_A\big)\,:\ \mathrm{Ext}^r_R(A,R)\otimes_R A \longrightarrow E\,,$$

(from which γ can be recovered as $\langle\gamma\lambda,a\rangle=\lambda(a)$ for all a in A, λ in $\mathrm{Hom}_R(A,E)$).

The foregoing takes on a slightly simpler form in case that R contains a copy of its residue class field, hence if it is in fact an augmented k-algebra. Then E can be chosen to be—and is always isomorphic to—the R-module $\mathrm{Hom}_k(R,k)$[10], so that $i\colon k\to E$ equals the k-dual of the projection $R\to k$. It follows that $\mathrm{Hom}_R(A,E)\cong\mathrm{Hom}_k(A,k)$ as R-modules and that the bilinear pairing $\langle\ ,\ \rangle_A$ takes its values already in k.

These statements extend trivially to the affine case in the following sense.

Assume given again an arbitrary commutative Gorenstein ring T of finite Krull dimension t. If $\mathcal{S}$ is some finite subset of $\mathrm{maxspec}(T)$, the set of all maximal ideals **66**
of T, the module $T/I(\mathcal{S})$[11] is semi-simple, namely by definition isomorphic to the product of all the residue class fields $k(\mathfrak{m})=T/\mathfrak{m}$, $\mathfrak{m}$ ranging over all the maximal ideals in $\mathcal{S}$.

An injective envelope of this module, denoted $E_{\mathcal{S}}$, is then the product of injective envelopes of the fields—and T-modules—$k(\mathfrak{m})$.

If now A is a T-module of finite length, it is supported at finitely many maximal primes, and the T-modules $\mathrm{Hom}_T(A,E_{\mathcal{S}})$, obtained from finite subsets of $\mathrm{maxspec}(T)$ which contain these finitely many maximal ideals, are all isomorphic.

Accordingly, we will denote—ambiguously—any of these modules by A'.

The local theory then yields for any dualizing T-module ω_T the existence of a natural T-linear isomorphism

$$A' \xrightarrow{\ \cong\ } \mathrm{Ext}^t_T(A,\omega_T)\,, \tag{7.6.4}$$

and the T-modules A, A' and $\mathrm{Ext}^t_T(A,\omega_T)$ are all of the same finite length.

7.7 Now we can finally formulate the application of the Duality Theorem (7.5.1) which seems to be most useful.

For this, assume given:

- $f\colon T\to S$ a duality morphism of finite type as in (7.6.2), of "virtual codimension" $d=\mathrm{projdim}_T\, f_*S$,
- ω_T a dualizing T-module, $\omega_S=f^!\omega_T[d]$ the associated dualizing module over S as in Remark (b) above,
- M a complex of left S-modules in $\mathrm{D^b}(S^{\mathrm{op}})$ as in (7.5.1),
- N an S-bimodule which is maximal Cohen–Macaulay as either a left or right S-module, considered as a T-module by restriction of scalars along f.

---

[10] Rather this should be the $\mathfrak{m}$-torsion submodule of $\mathrm{Hom}_k(R,k)$.

[11] $I(\mathcal{S})$ is the product of the maximal ideals in $\mathcal{S}$.

Then, summarising the foregoing discussion and substituting these data into (7.5.1), we get

PROPOSITION 7.7.1. (i) *Under the assumptions just made, the spectral sequence* $''E^{i,j}$ *for* M, N *and* $W = \omega_T$ *collapses at the* E_2*-level, leaving a single spectral sequence*

67 $$E_2^{i,j} = \mathrm{Ext}_T^i(\underline{\mathrm{Ext}}_{S^{\mathrm{op}}}^{-j}(M,N),\omega_T) \Longrightarrow \underline{\mathrm{Ext}}_S^{i+j-d-1}(M^*,\mathrm{Hom}_S(N,\omega_S))\,.$$

(ii) *Applying to the* E_2*-terms of this spectral sequence the trivial duality induced by* $\mathrm{Hom}_{S^{\mathrm{op}}}(-,S)$—(4.6.1) *and* (6.2.5) (2)—*it becomes*

$${}^*E_2^{i,j} = \mathrm{Ext}_T^i(\underline{\mathrm{Ext}}_S^{-j}(N^*,M^*),\omega_T) \Longrightarrow \underline{\mathrm{Ext}}_S^{i+j-d-1}(M^*,\mathrm{Hom}_S(N,\omega_S))\,.$$

PROOF. (i) Just observe that under the assumptions made,

$\mathrm{Ext}_T^i(f_*N,\omega_T) = \mathrm{Ext}_S^i(N,f^!\omega_T)$	by defintion of $(f_*,f^!)$
$= \mathrm{Ext}_S^{i-d}(N,\omega_S)$	by defintion of ω_S and (7.6.3), Remark (b) in (7.6),
$= 0 \quad$ for $i \neq d$,	N being MCM as a right S-module.

(ii) As N is by assumption also MCM as a *left* S-module, (6.2.5) 3 applies. (Remark that N finitely generated would be enough—at the expense of interpreting N^* again as $\mathrm{RHom}_{S^{\mathrm{op}}}(N,S)$.) □

In view of the local duality theory for commutative rings which we just recalled, the spectral sequences of the proposition degenerate if the Tate cohomology groups involved are of finite length.

For convenience (only), we will replace in the following M by M^* so that M will henceforth represent a complex of right S-modules in $\mathrm{D^b}(S)$.

Furthermore, if N is any S-bimodule, let us denote:

${}^*N = \mathrm{Hom}_{S^{\mathrm{op}}}(N,S)\,,$	the *left* S-dual of N, and
$N^* = \mathrm{Hom}_S(N,S)\,,$	the *right* S-dual of N.

Both these modules are naturally S-bimodules again, but only their underlying right module structure will be used in the sequel.

Also, the following terminology will shorten formulations:

DEFINITION 7.7.2. Let $f\colon T \to S$ be a homomorphism of rings satisfying (7.6.2).

(i) Two complexes of S-modules, Y in $\mathrm{D}^-(\mathrm{Mod}(S))$ and X in $\mathrm{D^b}(S)$, are (essentially) *stably transversal* (with respect to f) iff for any integer j, the T-module $\underline{\mathrm{Ext}}_S^j(X,Y)$ is of finite length.

68 (ii) The ring homomorphism f will be called *stably regular outside* $\mathcal{S}$ for a subset $\mathcal{S}$ of $\mathrm{Spec}(T)$ iff for all complexes X and Y as above, the groups $\underline{\mathrm{Ext}}_S(X,Y)$ are supported on $\mathcal{S}$.

Accordingly, f is *stably regular* at a prime $\mathfrak{p}$ in $\mathrm{Spec}(T)$ iff the localizations $\underline{\mathrm{Ext}}_S(X,Y)_{\mathfrak{p}}$ vanish for all X and Y.

DEFINITION 7.7.3. Let S be a ring which is strongly Gorenstein. Then S has only *isolated singularities* if there exists a homomorphism $f\colon T \to S$ which satisfies (7.6.2) and is stably regular outside a finite set of maximal ideals of T.

In this case, E_f will represent any injective envelope over T of $T/I(\mathcal{S})$, $\mathcal{S}$ some finite subset of maxspec(T) off which f is stably regular.

REMARK. The terminology should be justified by the following elementary observation (cf. also [**Aus86**]):

If $f\colon T \to S$ is a homomorphism as in (7.6.2), let $\mathfrak{p}$ be a prime ideal in T. Then $S_\mathfrak{p} = S \otimes_T T_\mathfrak{p}$ is a flat S-module (on both sides) and localization at $\mathfrak{p}$ defines trivially an exact functor from $\mathrm{D}(S)$ into $\mathrm{D}(S_\mathfrak{p})$. It is left as an exercise to establish that with S also $S_\mathfrak{p}$ is strongly Gorenstein and that Tate cohomology localizes:

$$\underline{\mathrm{Ext}}_S(X,Y) \otimes_T T_\mathfrak{p} = \underline{\mathrm{Ext}}_{S_\mathfrak{p}}(X_\mathfrak{p}, Y_\mathfrak{p})\,.$$

In particular, f is stably regular at $\mathfrak{p}$ if $S_\mathfrak{p}$ is a ring of finite global dimension.

If S is furthermore commutative, also the converse holds: Let $\mathfrak{q}$ be a prime in S lying over $\mathfrak{p}$ in T. If $S_\mathfrak{q}$ is not regular, consider the cyclic S-module $S/\mathfrak{q}$ whose localization at $\mathfrak{q}$ is the residue class field $k(\mathfrak{q})$ of $S_\mathfrak{q}$. Localising subsequently at $\mathfrak{p}$ and $\mathfrak{q}$ yields maps

$$\underline{\mathrm{Hom}}_S(S/\mathfrak{q}, S/\mathfrak{q}) \longrightarrow \underline{\mathrm{Hom}}_S(S/\mathfrak{q}, S/\mathfrak{q}) \otimes_T T_\mathfrak{p} \longrightarrow \underline{\mathrm{Hom}}_{S_\mathfrak{q}}(k(\mathfrak{q}), k(\mathfrak{q}))$$

under which the identity on $S/\mathfrak{q}$ is mapped onto the identity on $k(\mathfrak{q})$.

As $k(\mathfrak{q})$ is not of finite projective dimension, it represents not a zero-object in $\underline{\mathrm{mod}}(S_\mathfrak{q})$ by (2.1). Consequently, the localization of $\underline{\mathrm{Hom}}_S(S/\mathfrak{q}, S/\mathfrak{q})$ at $\mathfrak{p}$ is not zero and hence f is not stably regular at this prime of T.

It seems worth-while to record this as a separate

LEMMA 7.7.4. *Let S be a commutative Gorenstein ring of finite Krull di-* **69**
mension. Then for a prime $\mathfrak{p}$ in Spec(S) *the groups $\underline{\mathrm{Ext}}_S(X,Y)_\mathfrak{p}$ are zero for all (complexes of) S-modules Y (in $\mathrm{D}^-(\mathrm{Mod}(S))$) and X (in $\mathrm{D^b}(S)$) if and only if $S_\mathfrak{p}$ is a regular local ring.* □

Coming back to the promised application of the Duality Theorem, let us recall that $t = \dim T$ and $s = \operatorname{vdim} S = t - d$ by (7.6.3) (iii)[12]. Then we have:[13]

THEOREM 7.7.5 (The Duality Theorem for isolated singularities). *Assume again given a homomorphism of rings $f\colon T \to S$ which satisfies* (7.6.2), *ω_T and ω_S dualizing modules over T and S respectively as in* (7.7.1), *M a complex of S-modules in $\mathrm{D^b}(S)$ and N an S-bimodule which is MCM on both sides.*

(i) *If M and the underlying right S-module of N are stably transversal with respect to f, then for any integer j there exists a natural isomorphism of T-modules*

$$\gamma^j_{M,N}\colon \underline{\mathrm{Ext}}_S^{-j}({}^*N, M)' \xrightarrow{\ \cong\ } \underline{\mathrm{Ext}}_S^{j+s-1}(M, \omega_S \otimes_S N^*)\,.$$

(ii) *Let L be a second complex in $\mathrm{D^b}(S)$, also stably transversal to the underlying right S-module of N. If then α is a class in $\underline{\mathrm{Ext}}_S^k(M,L)$, one has a commutative diagram of T-linear maps*

$$\begin{array}{ccc}
\underline{\mathrm{Ext}}_S^{-j}({}^*N, L)' & \xrightarrow[\cong]{\gamma^j_{L,N}} & \underline{\mathrm{Ext}}_S^{j+s-1}(L, \omega_S \otimes_S N^*) \\
{\scriptstyle (\alpha\circ -)'}\big\downarrow & & \big\downarrow{\scriptstyle (-\circ\alpha)} \\
\underline{\mathrm{Ext}}_S^{-k-j}({}^*N, M)' & \xrightarrow[\cong]{\gamma^{k+j}_{M,N}} & \underline{\mathrm{Ext}}_S^{k+j+s-1}(M, \omega_S \otimes_S N^*)\,,
\end{array}$$

---

[12] See Remark A.7.1.

[13] The notation $(-)'$ is from the paragraph before (7.6.4).

where the vertical maps are given by the Yoneda products with α.

(iii) *Assume that* f *is stably regular outside a finite set of maximal ideals of* T—*so that in particular* S *has only isolated singularities. Then the inverse of* $\gamma^j_{M,N}$ *yields a non-degenerate* T*-bilinear pairing*

$$_M\langle\ ,\ \rangle^j_N\colon \underline{\mathrm{Ext}}_S^{j+s-1}(M,\omega_S\otimes_S N^*)\otimes_T \underline{\mathrm{Ext}}_S^{-j}({}^*N,M)\longrightarrow E_f\,,$$

associative in the sense that for any class α *in* $\underline{\mathrm{Ext}}_S^k(M,L)$ *as above,* x *in* $\underline{\mathrm{Ext}}_S^{j+s-1}(L,\omega_S\otimes_S N^*)$ *and* y *in* $\underline{\mathrm{Ext}}_S^{-k-j}({}^*N,M)$:

$$_M\langle x\circ\alpha,y\rangle^{k+j}_N = {}_L\langle x,\alpha\circ y\rangle^j_N\,.$$

70 (iv) *Restricting* (iii) *to the special case* $j=0$ *and* $M={}^*N$, *it follows for any* S*-bimodule* N, *which is MCM on both sides, the existence of a* T*-linear trace map:*

$$\tau_N := {}_{^*N}\langle\ ,\mathrm{id}_{^*N}\rangle^0_N\colon \underline{\mathrm{Ext}}_S^{s-1}({}^*N,\omega_S\otimes_S N^*)\longrightarrow E_f\,,$$

such that in general the pairing in (iii) *can be obtained by composing this trace with the Yoneda product:*

$$_M\langle x,y\rangle^j_N = \tau_N(x\circ y)$$

for all x *in* $\underline{\mathrm{Ext}}_S^{j+s-1}(M,\omega_S\otimes_S N^*)$ *and* y *in* $\underline{\mathrm{Ext}}_S^{-j}({}^*N,M)$.

PROOF. The proof consists just of a reformulation of (7.7.1) in these special circumstances.

Consider the spectral sequence ${}^*E_2^{i,j}\implies \mathbf{E}^{i+j}$ of (7.7.1) (ii) for the arguments M^* (instead of M), N and ω_T, ω_S as stated. Then ${}^*E_2^{i,j}=\mathrm{Ext}^i_T(\underline{\mathrm{Ext}}_S^{-j}({}^*N,M),\omega_T)$, as $M=M^{**}$ in $\mathrm{D^b}(S)$. Moreover M being stably transversal to ${}^*N$ with respect to f by assumption, the local duality theory shows that ${}^*E_2^{i,j}=0$ for $i\neq t$ and that $\underline{\mathrm{Ext}}_S^{-j}({}^*N,M)'$ is naturally isomorphic to ${}^*E_2^{t,j}$.

Consequently, the spectral sequence degenerates at the E_2-level, the edge homomorphisms $e_2^{t,j}\colon {}^*E_2^{t,j}\to\mathbf{E}^{t+j}$ are defined and yield isomorphisms of T-modules

$$\underline{\mathrm{Ext}}_S^{-j}({}^*N,M)'\longrightarrow {}^*E_2^{t,j}\xrightarrow{\ e_2^{t,j}\ }\mathbf{E}^{t+j}\,.$$

Furthermore, N being MCM as a right S-module, there is a natural isomorphism—of S-bimodules even—from $\omega_S\otimes_S N^*$ into $\mathrm{Hom}_S(N,\omega_S)$, whose inverse induces an isomorphism of left S-modules

$$\mathbf{E}^{t+j}=\underline{\mathrm{Ext}}_S^{t+j-d-1}(M,\mathrm{Hom}_S(N,\omega_S))\xrightarrow{\ \cong\ }\underline{\mathrm{Ext}}_S^{j+s-1}(M,\omega_S\otimes_S N^*)$$

in view of $s=t-d$. Composing these isomorphisms yields the desired maps $\gamma^j_{M,N}$ and proves (i).

Assertion (ii) then follows from (7.5.4) by applying the duality $\mathrm{Hom}_{S^{\mathrm{op}}}(-,S)$ to the contravariant arguments of E_2, taking into account the behavior of Yoneda products with respect to this duality, (7.1.2).

Given (i) and (ii), (iii) and (iv) are well-known and easily established consequences. □

71 REMARK. The trace maps exhibited in (iv) should be of paramount importance—and hard to come by.

This will be supported by Chapter 10 which indicates furthermore that an explicit determination of these traces may require a "calculus of residues" as in algebraic geometry.[14]

7.8 Noether different and support of Tate cohomology. Before treating concrete examples, we will give another result on the "universal" support of Tate cohomology, supplementing in the commutative case Lemma 7.7.4. Informally stated, the result is:

Assume that $f\colon T \to S$ is a ring homomorphism from a commutative ring of finite Krull dimension and of *finite global dimension* into the center of a ring S which is strongly Gorenstein such that f_*S, the underlying T-module of S, is finitely generated projective. Then:

The Noether different of f annihilates all the
Tate cohomology groups $\underline{\mathrm{Ext}}_S(-,-)$ over S.

To make this statement precise, we first need to define the Noether different in the general non-commutative setting, naturally extending the usual notion in the commutative case as introduced by E. Noether and investigated more closely by R. Berger, [**Ber61**]. Such extensions have been studied by several authors; see [**Fos66**] for further references.

To begin with, assume given an arbitrary commutative ring A and a ring homomorphism $g\colon A \to B$ where B is an associative ring with unit as usual, such that B becomes a (central) A-algebra with respect to g. Then, recall that

$$B^{\mathrm{e}}_A := B \otimes_A B^{\mathrm{op}}$$

is the *enveloping algebra* of g—or, less precisely of B over A.

Deviating from our general convention, we will only deal with left modules over enveloping algebras. In particular, B becomes a left B^{e}_A-module in the usual way by

$$\lambda\colon B^{\mathrm{e}}_A \otimes_A B \to B\,, \quad \lambda(b' \otimes b'' \otimes b) = b'bb''\,.$$

The induced map

$$\mu\colon B^{\mathrm{e}}_A \to B\,, \quad \mu(b' \otimes b'') = \lambda(b' \otimes b'' \otimes 1) = b'b''$$

is then a homomorphism of left B^{e}_A-modules, also called the natural *left augmentation* of B^{e}_A. The kernel of μ is denoted $I_{B/A}$ and accordingly called the (left) **72**
augmentation ideal of B^{e}_A. It represents naturally the two-sided A-linear derivations on B, [**Bou70**, III.132], and is generated as a left B^{e}_A-ideal—or even as a left B-module—by the image of the universal derivation $d_{B/A}\colon B \to I_{B/A}$ of B over A, given by

$$d_{B/A}(b) = b \otimes 1 - 1 \otimes b\,.$$

It follows that the right annihilator $\mathrm{ann}(I_{B/A})_{B^{\mathrm{e}}_A}$ is mapped by μ into the center $\mathrm{Z}(B)$ of B: An element w in B^{e}_A annihilates the left ideal $I_{B/A}$ from the right if and only if $d_{B/A}(b) \cdot w = 0$ for all b in B. Hence

$$0 = \mu(d_{B/A}(b)w) = b \cdot \mu(w) - \mu(w) \cdot b$$

for all b as required.

[14]This prescient remark has been validated by Murfet; see (C.4.2).

DEFINITION 7.8.1. The (classical) *Noether different* $\vartheta_N^0(g)$ of the A-algebra $g\colon A \to B$ is that ideal of the center $\mathrm{Z}(B)$ of B which is given by

$$\vartheta_N^0(g) := \mu(\mathrm{ann}(I_{B/A})_{B_A^{\mathrm{e}}}) .$$

This definition is E. Noether's original one in the commutative case, extended by M. Auslander–O. Goldman to the non-commutative case. The following homological description seems due to D. G. Higman; see once again [**Ber61**] and [**Fos66**].

For our purposes, it is more convenient to interpret the Noether different in homological terms:

- $\mathrm{Z}(B)$ may be identified with $\mathrm{Hom}_{B_A^{\mathrm{e}}}(B, B)$, the ring of B-bimodule endomorphisms of B, by associating to such an endomorphism Θ the element $\Theta(1)$.
- $\mathrm{ann}(I_{B/A})_{B_A^{\mathrm{e}}}$ can be identified with $\mathrm{Hom}_{B_A^{\mathrm{e}}}(B, B_A^{\mathrm{e}})$ by associating to a B-bimodule homomorphism Θ from B into B_A^{e} again the element $\Theta(1)$. Conversely, an element w in the right annihilator of $I_{B/A}$ defines the left B_A^{e}-homomorphism $\Theta_w(b) = (b \otimes 1)w = (1 \otimes b)w$.
- $\mathrm{Z}(B) = \mathrm{Hom}_{B_A^{\mathrm{e}}}(B, B)$ acts naturally from the right on $\mathrm{Hom}_{B_A^{\mathrm{e}}}(B, -)$ by composition, so that the induced map $\mathrm{Hom}_{B_A^{\mathrm{e}}}(B, \mu)$ is naturally $\mathrm{Z}(B)$-linear on the right. Furthermore, evaluation at the unit of B identifies this map with the restriction of μ to $\mathrm{ann}(I_{B/A})_{B_A^{\mathrm{e}}}$:

73

$$\begin{array}{ccccc}
\mathrm{Hom}_{B_A^{\mathrm{e}}}(B, \mu) : & \mathrm{Hom}_{B_A^{\mathrm{e}}}(B, B_A^{\mathrm{e}}) & \longrightarrow & \mathrm{Hom}_{B_A^{\mathrm{e}}}(B, B) \\
& \downarrow{\scriptstyle\cong} & & \downarrow{\scriptstyle\cong} \\
\mu|_{\mathrm{ann}(I_{B/A})_{B_A^{\mathrm{e}}}} : & \mathrm{ann}(I_{B/A})_{B_A^{\mathrm{e}}} & \longrightarrow & \mathrm{Z}(B) .
\end{array}$$

From this description of the Noether different one has immediately

PROPOSITION 7.8.2. *Let $g\colon A \to B$ be an A-algebra as above. Then $\vartheta_N^0(g) = \mathrm{Z}(B)$ if and only if B is projective as a left module over the enveloping algebra B_A^{e}.*

If B is commutative and g_*B, the underlying A-module of B, is projective, then $\vartheta_N^0(g) = B$ holds if and only if $g\colon A \to B$ is an étale A-algebra.

PROOF. By the above, $\vartheta_N^0(g) = \mathrm{Z}(B)$ iff $\mathrm{Hom}_{B_A^{\mathrm{e}}}(B, \mu)$ is surjective. This happens iff the short exact sequence of left B_A^{e}-modules

$$0 \longrightarrow I_{B/A} \longrightarrow B_A^{\mathrm{e}} \xrightarrow{\ \mu\ } B \longrightarrow 0$$

splits, hence iff B is B_A^{e}-projective as a left module. The characterization of a (commutative) finite and étale algebra by its Noether different then just restates [**Gro67**, 18.3.1.(ii)]. □

The significance of Noether differents for stable homological algebra—hence in particular for Tate cohomology—is due to the following elementary observation which "derives" in a straightforward manner the treatment of bimodules as given by H. Bass in [**Bas68**, II.2]:

Let X be a bounded above complex of B-bimodules whose underlying complex of A-bimodules is "symmetric" in the sense that its left and right module structures over the commutative ring A are the "same". Then X corresponds bi-uniquely to a complex of left B_A^{e}-modules and such complexes represent hence all the objects of the derived category $\mathrm{D}^-(\mathrm{Mod}(B_A^{\mathrm{e}}))$ as defined in (7.4).

On the other hand, taking tensor products with respect to the underlying *left* B-module structure of X and remembering the *right* B-module structure of X afterwards, defines an *exact* functor

$$\mathbf{h}(-,X) := - \otimes^{\mathsf{L}}_B X \colon \mathrm{D}^-(\mathrm{Mod}(B)) \longrightarrow \mathrm{D}^-(\mathrm{Mod}(B))\,.$$ 74

This functor commutes *canonically* with translation on $\mathrm{D}^-(\mathrm{Mod}(B))$ by [**Bou80**, 61 Rem.(2)] and a quasi-isomorphism of complexes of B-bimodules yields a natural equivalence of functors.

Leaving the pain of defining properly the "category of additive and exact endofunctors on $\mathrm{D}^-(\mathrm{Mod}(B))$" to the reader, $\mathbf{h}(-,-)$ can hence be interpreted as a functor from $\mathrm{D}^-(\mathrm{Mod}(B^{\mathrm{e}}_A))$ into that category.

Look at the following examples:

- If $X = B$, considered as a (complex of) left B^{e}_A-module(s), $\mathbf{h}(-,B)$ is isomorphic to the identity functor on $\mathrm{D}^-(\mathrm{Mod}(B))$.
- If $X = B^{\mathrm{e}}_A$ considered as a (complex of) left B-module(s), the associativity formula for the derived tensor product
$$- \otimes^{\mathsf{L}}_B (B \otimes^{\mathsf{L}}_A B^{\mathrm{op}}) \cong - \otimes^{\mathsf{L}}_A B^{\mathrm{op}}$$
shows that
$$\mathbf{h}(-,B^{\mathrm{e}}_A) \cong g^* g_* \colon \mathrm{D}^-(\mathrm{Mod}(B)) \longrightarrow \mathrm{D}^-(\mathrm{Mod}(A)) \longrightarrow \mathrm{D}^-(\mathrm{Mod}(B))\,,$$
as soon as the natural augmentation onto the 0th-homology:
$$B \otimes^{\mathsf{L}}_A B^{\mathrm{op}} \longrightarrow \mathrm{H}^0(B \otimes^{\mathsf{L}}_A B^{\mathrm{op}}) = B \otimes_A B^{\mathrm{op}} = B^{\mathrm{e}}_A$$
is a quasi-isomorphism, which is equivalent to
$$\mathrm{H}^{-i}(B \otimes^{\mathsf{L}}_A B^{\mathrm{op}}) = \mathrm{Tor}^A_i(B,B) = 0 \quad \text{for all } i \neq 0\,.$$
(Remark that $B = B^{\mathrm{op}}$ as A-modules, $g(A)$ being central in B.)
- The morphism μ of left B^{e}_A-modules yields a morphism of functors $\mathbf{h}(-,\mu)$, and if $\mathrm{Tor}^A_i(B,B) = 0$ for $i \neq 0$, it becomes identified with the *adjunction co-unit* $g^*g_* \xrightarrow{\varepsilon} \mathrm{id}$.

  In case that B is projective as an A-module, one may forget about deriving tensor products and the identification just states the obvious:
$$\begin{aligned} x(b'b'') = (\mathrm{id}_X \otimes \mu)(x \otimes_B b' \otimes_A b'') &= (\mathrm{id}_X \otimes \mu)(xb' \otimes_B 1 \otimes_A b'') \\ &= \varepsilon(X)(xb' \otimes_A b'') = (xb')b'' \end{aligned}$$
for all elements x in X and b', b'' in B.
- Finally, if z is an element of the center of B, considered as a B-bimodule endomorphism of B as above, $\mathbf{h}(-,z)$ yields a natural transformation of the identity functor on $\mathrm{D}^-(\mathrm{Mod}(B))$—and, generalising [**Bas68**, II.2.1.Prop.], one may prove that all such transformations are obtained this way.

Consequently, if z is already an element of the Noether different $\vartheta^0_N(g)$, ob- 75
tained as $\mu \circ w$ for some $w \colon B \to B^{\mathrm{e}}_A$, we may interpret these data as natural transformations of functors:

$$\begin{array}{ccccccc} \mathbf{h}(-,z): & \mathbf{h}(-,B) & \xrightarrow{\mathbf{h}(-,w)} & \mathbf{h}(-,B^{\mathrm{e}}_A) & \xrightarrow{\mathbf{h}(-,\mu)} & \mathbf{h}(-,B) \\ & \downarrow\cong & & \downarrow\cong & & \downarrow\cong \\ \rho(z)\ : & \mathrm{id}_{\mathrm{D}^-(\mathrm{Mod}(B))} & \longrightarrow & g^*g_* & \xrightarrow{\varepsilon} & \mathrm{id}_{\mathrm{D}^-(\mathrm{Mod}(B))}\,, \end{array}$$

where $\rho(z)$ denotes "right multiplication with z" regarded as a natural transformation of the identity functor on $\mathrm{D}^{-}(\mathrm{Mod}(B))$.

Hence, *if*:

- B is noetherian and g_*B, the underlying A-module, is finitely generated projective (which ensures in particular $\mathrm{Tor}_i^A(B,B) = 0$ for $i \neq 0$),
- A is a ring of finite global dimension, and
- X is a complex in $\mathrm{D}^{\mathrm{b}}(B)$—see (0.2) for notation,

then:

- Multiplication with an element z in the Noether different $\vartheta_N^0(g)$ from the right on X factors through a *perfect* complex, namely g^*g_*X.

Specialising to the case of Gorenstein rings, this yields the result announced:

THEOREM 7.8.3. *Let $f\colon T \to S$ be a homomorphism of rings where*

- *S is strongly Gorenstein and f_*S is finitely generated projective,*
- *T is a commutative ring of finite Krull dimension such that all localizations $T_{\mathfrak{p}}$, $\mathfrak{p}$ a prime in* $\mathrm{Spec}(T)$, *are regular local rings,*[15]
- *and, as usual, $f(T)$ is contained in the center* $\mathrm{Z}(S)$ *of S.*

Then the natural action of $\mathrm{Z}(S)$, *on* $\underline{\mathrm{Ext}}_S(X,Y)$ *annihilates the Noether different $\vartheta_N^0(f)$ for all Y in* $\mathrm{D}^{-}(\mathrm{Mod}(S))$ *and X in* $\mathrm{D}^{\mathrm{b}}(S)$.[16]

PROOF. Simply remark that the hypotheses on T guarantee that its global dimension is finite. □

REMARK. Classically, in the commutative case, one rather studies the *Kähler or Dedekind different*, denoted here $\vartheta_K(-)$ or $\vartheta_D(-)$ respectively.

The relations for a commutative ring S and a morphism $f\colon T \to S$ as in the theorem are the following:

76 (For the definitions and properties of $\vartheta_{K/D}(f)$ the reader may consult [**Ber61**] or [**HK71**] and the literature cited there.)

(i) If $I_{S/T}$ is finitely generated as a left S_T^{e}-ideal, the Noether different $\vartheta_N^0(f)$ contains the Kähler different $\vartheta_K(f)$ [**Ber61**, II.Satz 4].

(ii) If $I_{S/T}/(I_{S/T})^2$ is finitely generated as an S-module, say by m elements, then the Kähler different contains the m-th power of the Noether different, [**Ber61**, II.Satz 3].

(iii) If f is *generically unramified* in the sense that for K the field of fractions of T the extension $f \otimes_T K\colon K \to S \otimes_T K$ is separable, then Noether and Dedekind different of f coincide, [**Ber61**, III.Satz 7].

(iv) If f is a *complete intersection* morphism in the sense that in addition to f being finite and flat all fibres are—necessarily artinian—complete intersection algebras, then all three differents are the same; E. Kunz, [**Kun68**].

The properties (i)–(iii) do not depend on S being Gorenstein, the morphism f being finite and flat—which implies S Cohen–Macaulay—is enough. (In (iv), the Gorenstein property is implied by the other assumptions.) A special property of Gorenstein rings is nevertheless:

(v) The Dedekind different is a *principal ideal* in S, [**HK71**, 7.25].

---

[15] It is enough to assume that T is regular, for that already ensures that f_*X is perfect; see Remark A.7.2.

[16] In other words, the Noether different $\vartheta_N^0(f)$ annihilates $\underline{\mathrm{Ext}}_S(X,Y)$.

Does (v) still hold for the Noether different of a not necessarily commutative but still strongly Gorenstein ring S?

Another remark concerns the tensor product used to define the enveloping algebra. It should depend on the *category of algebras* one is working with and not just on the ring theoretic properties.

For example, treating *analytic* commutative local algebras over some field, hence quotients of rings of convergent or formal power series, the tensor product should be completed to obtain the "correct" analytic different.

Similarly, in the *graded commutative* case, the tensor product, could be replaced by the graded tensor product (with respect to the given commutation factor), see [**Bou70**, III.47], to obtain a graded version of the different.

All this shows that Noether differents are rather a general homological tool not necessarily bound to the theory of Gorenstein rings.

We conclude this section with three, fairly well-known examples and an application to (commutative) complete intersections.

The notations will be as before, $g\colon A \to B$ denoting an A-algebra with envelop- **77**
ing algebra B^{e}_A.

EXAMPLE 7.8.4 (Group algebras). Let A be an arbitrary commutative ring and G a finite group. Set $B = A[G]$, the group algebra of G over A, and let $\iota\colon A \to A[G]$ be the canonical map.

Then $\iota_* B$ is freely generated over A by the elements of G and the ideal $I_{B/A}$ in B^{e}_A is generated by the elements

$$d_{B/A}(h) = h \otimes 1 - 1 \otimes h\,, \quad h \text{ any element in } G\,.$$

It follows that the element

$$w = \sum_{g\in G} g \otimes g^{-1}$$

annihilates $I_{B/A}$ from the right:

$$d_{B/A}(h)w = \sum_{g\in G}(hg \otimes g^{-1} - g \otimes g^{-1}h) = 0$$

and hence, as

$$\mu(w) = \sum_{g\in G} 1 = [G:1]\,,$$

the *order of* G is an element of the Noether different. (Compare for example [**CE99**, XII.2.2.–2.5.].)

EXAMPLE 7.8.5 (Simple extensions). Let A again be arbitrary commutative, x a variable and $f(x)$ a polynomial with coefficients in A, $B = A[x]/(f(x))$ the simple extension defined by $f(x)$.

Then B^{e}_A is isomorphic to $A[x,y]/(f(x), f(y))$, y a second variable, and $\mu\colon B^{\mathrm{e}}_A \to B$ becomes the substitution "$x = y$".

Accordingly, $x - y$ generates $I_{B/A}$ and one may write

$$f(x) - f(y) = (x-y)g(x,y) \quad \text{for some } g(x,y) \text{ in } A[x,y]\,.$$

Evaluating this equation in the quotient ring B^{e}_A, it follows that the class of the "difference quotient" $g(x,y)$ in B^{e}_A annihilates the ideal $I_{B/A}$, whence

$$\mu(g(x,y)) = g(x,x) = f'(x)$$

shows that the *derivative* $f'(x)$ belongs to the Noether different of B over A. (In fact, it generates the Noether different [**Ber61**, II.1.Kor.3].)

78 EXAMPLE 7.8.6 (Exterior algebras). Let A be a commutative ring, V a free A-module of rank n. Consider $B = \bigwedge_A V$, the exterior algebra of V over A and the natural map $g\colon A \to B$.

If $v_1, \ldots, v_n$ constitutes an A-basis of V, the ideal $I_{B/A}$ in B^{e}_A is generated by the elements

$$d_{B/A}(v_i) = v_i \otimes 1 - 1 \otimes v_i\,, \quad i = 1, \ldots, n\,.$$

Then the product

$$w = (v_1 \otimes 1 + 1 \otimes v_1) \cdots (v_n \otimes 1 + 1 \otimes v_n)$$

is certainly in the right annihilator of $I_{B/A}$ and its image under μ,

$$\mu(w) = 2^n v_1 \cdots v_n\,,$$

is easily seen to generate the Noether different of g.

If one takes the *graded enveloping algebra* of B over A instead, which is $B \otimes_A B$ with the multiplication

$$(x_1 \otimes y_1)(x_2 \otimes y_2) = (-1)^{\deg y_1 \cdot \deg x_2}(x_1 x_2 \otimes y_1 y_2)$$

for homogeneous elements x_2, y_1, then one verifies that the corresponding graded Noether different is the zero ideal.

In both examples (7.8.5), 7.8.6 the information obtained is not always optimal: Take A a field of characteriztic two, $B = A[x]/(x^2)$ considered either as a simple extension or the exterior algebra of a free module of rank one. Then B is strongly Gorenstein and Theorem 7.8.3 applies to the canonical map $g\colon A \to B$. The various Noether differents mentioned above will all be zero but the element x in B annihilates all Tate cohomology groups over B as follows from (4.4.1).

Apart from this rather trivial example, it follows already from the general theory above that Noether differents are not always sufficient to describe precisely the "universal" support of Tate cohomology, even for commutative (Gorenstein) algebras S over a field K which allow "many" finite flat morphisms from regular rings T into it: By (7.8.2), the Noether different yields the non-étale locus of such a finite flat map, hence the union of all such Noether differents will describe (maximally) the non-smooth locus of S over K whereas the Tate cohomology groups are supported only on the non-regular locus of S by (7.7.4). (For more examples along these lines, see [**BEH87**].)

79 Extending (7.8.5) or directly using Remark (v) above, we get the following result on the Tate cohomology of complete intersections in the equi-characteriztic case:

Let k be a field, and denote by $P = k\langle x_1, \ldots, x_n\rangle$ either

- the localization of the polynomial ring over k in n variables at the maximal ideal $\mathfrak{m} = (x_1, \ldots, x_n)$ or
- the ring of formal (or convergent) power series over a (complete non-trivially valuated) field k in n variables.

Assume given a sequence $\underline{f} = (f_1, \ldots, f_m)$ of elements in the unique maximal ideal $\mathfrak{m} = (x_1, \ldots, x_n)$ and let $J(\underline{f})$ denote the corresponding *Jacobian ideal* generated by all maximal minors of the Jacobian matrix of $\underline{f}$ with respect to the chosen coordinate-functions $x_1, \dots, x_n$, whose entries are hence given by the partial derivatives $\partial f_j/\partial x_i$.

With these notations we have:

COROLLARY 7.8.7. [17] *If $\underline{f} = (f_1, \dots, f_m)$ constitutes a regular sequence in $\mathfrak{m} = (x_1, \dots, x_n)$, the quotient ring $R = P/\underline{f}P$ is a complete intersection ring, hence Gorenstein, of dimension $n - m$.*

The Jacobian ideal $J(\underline{f})$ then annihilates all Tate cohomology groups over R which become consequently modules over the Jacobian ring $\overline{R} = R/J(\underline{f})R$ of R in a natural way.

PROOF. It needs only to be remarked that a composition

$$k\langle x_{i_1}, \dots, x_{i_{n-m}}\rangle \xrightarrow{\text{incl.}} P \xrightarrow{\text{proj.}} R$$

yields a finite flat homomorphism of rings if the corresponding minor

$$\det(\partial f_j/\partial x_i)^{j=1,\dots,m}_{i\neq i_1,\dots,i_{n-m}}$$

of the Jacobian matrix is not zero. In this case, the associated ("analytic"!) Kähler different—which equals here the (analytic) Noether different in accordance with Remark (iv) above—is generated by that minor, see [**Kun68**, Sätze 1 & 2]. □

[17]For recent developments see C.6.

CHAPTER 8

First examples

First, let us consider those cases where S is of small virtual dimension. **80**

8.1 Quasi-Frobenius rings. Let S be an artinian ring which is injective as a right module over itself. Then it is also injective on the left, [**CR81**, p. 135], hence S is strongly Gorenstein of virtual dimension zero. Such rings are also called *quasi-Frobenius*—which motivated Heller's definition of a Frobenius category, (4.8).

For S quasi-Frobenius, obviously all finitely generated S-modules are maximal Cohen–Macaulay, whence $\underline{\mathrm{D}}^{\mathrm{b}}(S) = \underline{\mathrm{MCM}}(S) = \underline{\mathrm{mod}}(S)$. Furthermore, complete resolutions are simply obtained by connecting a projective with an injective resolution of the given module.

Let $\mathrm{rad}(S)$ denote the *radical* of S, $\mathrm{soc}(S)$ its *socle*.

PROPOSITION 8.1.1 (cf. [**CR81**, 6.28ff], for example). *Let S be a ring which is quasi-Frobenius. Then*

(i) $\mathrm{soc}(S)$ *is a two-sided ideal of S, canonically isomorphic to the S-dual of* $S/\mathrm{rad}(S)$.

(ii) *A finitely generated right S-module M contains a projective (or injective) direct summand if and only if* $M \cdot \mathrm{soc}(S) \neq 0$. □

This shows, by (2.1), that the isomorphism classes of objects in $\underline{\mathrm{MCM}}(S)$ are in bijective correspondence with the isomorphism classes of finitely generated right $\overline{S} = S/\mathrm{soc}(S)$-modules. Furthermore, supplementing (7.7.4) and (7.8.3) in this case, it follows from the proposition:

COROLLARY 8.1.2. *For any two (complexes of) S-modules M and N (in $\mathrm{D}^{\mathrm{b}}(S)$) over a quasi-Frobenius ring S, the Tate cohomology groups $\underline{\mathrm{Ext}}_S(M, N)$ are annihilated by all central elements in the socle of S.* □

In case of quasi-Frobenius rings, the "canonical octahedron" of a module, (5.3.1), degenerates, the functor $\mathbf{M}$ becomes the identity and $\mathbf{H}$ may be chosen to associate to a module its injective envelope. Hence, in case of quasi-Frobenius **81**
rings, the theory developed here reduces essentially to the representation theory of artinian rings which are self-injective.

Before treating more specific (graded) examples of such rings, where Tate cohomology admits a rather concrete geometric interpretation, we want to mention some more classes of rings which are strongly Gorenstein.

For example

8.2 Integral representations of finite groups. Let G be a finite group, $S = \mathbb{Z}[G]$ its *integral group ring*. Then S is strongly Gorenstein in the terminology adopted here and its injective dimension equals one. The natural ring homomorphism from $\mathbb{Z}$ into S satisfies the conditions (7.6.2). It follows that an S-module

M—or equivalently, an integral representation of G—is maximal Cohen–Macaulay over S if and only if the underlying $\mathbb{Z}$-module is MCM over the integers, which means nothing but M being $\mathbb{Z}$-free of finite rank in view of the structure theory of modules over principal ideal domains, for example. Hence, one indeed regains the classical theory of complete resolutions as initiated by J. Tate.

The (original) Tate cohomology of G with values in an integral representation M, usually denoted $\widehat{\mathrm{H}}(G, M)$, is accordingly nothing but $\underline{\mathrm{Ext}}_{\mathbb{Z}[G]}(\mathbb{Z}, M)$, with $\mathbb{Z}$ considered as a trivial G-module.

Classically, most of the results on Tate cohomology are deduced using the cocommutative Hopf-algebra structure on $\mathbb{Z}[G]$. To point out the connections, let us make the following remarks: Assume given two integral representations M and N of G, with M being MCM. Then, using the co-multiplication and inversion on the group ring, $\mathrm{Hom}_{\mathbb{Z}}(M, N)$ becomes in a natural way a (right) $\mathbb{Z}[G]$-module again and one easily gets that

$$\widehat{\mathrm{H}}^i(G, \mathrm{Hom}_{\mathbb{Z}}(M, N)) = \underline{\mathrm{Ext}}^i_{\mathbb{Z}[G]}(M, N) \tag{8.2.1}$$

for all integers i and functorially in both M and N.

Furthermore, the duality theory for Tate cohomology in this case is usually developed using cup-products induced from the Hopf-algebra structure. But, by the various uniqueness theorems for homological products, (or here also simply by [**Bou80**, 201, Exer. 10(c)]), these products are known to be the "same" as the Yoneda products up to a sign (remember that we deal exclusively with *right*
82 modules!). It follows then easily—for example by invoking the various uniqueness theorems once again—that the duality theorems obtained here reduce to the classical statements.

An immediate generalisation of integral representation theory that is still covered is obtained by replacing $\mathbb{Z}$ with any Dedekind ring. The resulting group ring is again strongly Gorenstein of virtual dimension one and maximal Cohen–Macaulay modules over it correspond to *lattices.*

Finally, as far as group representations are concerned, let us mention the following two cases which display features rather similar to those described here but which are not covered. Maybe, these constitute a good testing ground for a more general theory of—not yet defined—"Gorenstein categories"[1].

(8.2.2) *Pro-finite Poincaré groups.* Considering the results on such groups—see [**Ser02**] for example—and in particular, the existing duality theory for them, (loc. cit.; app.), it seems reasonable to expect that large parts of the foregoing theory can be extended to cover these groups.

Even more promising, consider

(8.2.3) *Groups of finite virtual cohomological dimension.* For these groups, F. T. Farrell introduced a cohomology theory generalising naturally the Tate cohomology for finite groups. If the group is furthermore a "virtual duality group"—see [**Bro82**, VIII & X] for details—all the necessary ingredients like complete resolutions and the duality theory are available and developed in complete analogy to the

[1]See Remark (C.4.4) for subsequent developments.

classical case. This indicates that our finiteness assumptions—S noetherian and MCMs being finitely generated—should be largely redundant.[2]

Coming back to the classical case of finite groups some questions even for the theory developed here impose themselves. For example, what are the analogues for strongly Gorenstein rings of more sophisticated results like the theorem of Tate–Nakayama in class field theory and what are the geometric interpretations in case of isolated Gorenstein singularities?

For a simple example along these lines, consider the generalisation of "Herbrand quotients" to hypersurface singularities and their suspected geometric meaning outlined in Chapter 10 below.

83

8.3 Higher dimensional Gorenstein rings. Here we give two criteria which enable one to recognize strongly Gorenstein rings of arbitrarily large virtual dimension. The first one shows that strongly Gorenstein rings may be obtained by "deforming" those of small virtual dimension:

PROPOSITION 8.3.1. *Assume given a commutative regular local ring T with residue field k. If then $f\colon T \to S$ is a finite and flat T-algebra—see* (7.6)*—such that the ring $S_0 = S \otimes_T k$, the "geometric fibre" of f, is quasi-Frobenius, then S is strongly Gorenstein and its virtual dimension equals* $\dim T$.

Furthermore, f satisfies the conditions (7.6.2) *and consequently a finitely generated S-module is maximal Cohen–Macaulay if and only if the underlying T-module is free of finite rank.* □

The proof, which uses the well-known behavior of injective dimension with respect to quotients by central non-zero-divisors, is left to the reader.

It follows by the usual arguments that being strongly Gorenstein is a *generic property* of finite flat maps:

COROLLARY 8.3.2. *Let T be a commutative ring of finite Krull dimension all of whose localizations (at maximal ideals) are regular local.*

If then $f\colon T \to S$ is a finite flat T-algebra, $\mathfrak{m}$ a maximal ideal of T such that the corresponding fibre $S(\mathfrak{m}) = S \otimes_T T/\mathfrak{m}$ is a quasi-Frobenius ring, then there exists a Zariski-open[3] *neighbhorhood U of $\mathfrak{m}$ in* $\operatorname{Spec}(T)$, *such that $S_U = S \otimes_T T_U$ is strongly Gorenstein.*

In particular, S itself is strongly Gorenstein if and only if all geometric fibres over maximal ideals in T are quasi-Frobenius.[4] □

A simple example where the foregoing applies is given by group rings $T[G]$, with G a finite group, T a commutative ring satisfying the hypotheses of the corollary. Remark that then (7.8.4) implies the well-known fact that $T[G]$ is of finite global dimension as soon as $[G:1]$, the order of G, is invertible in T.

Similarly, coming back to the example (7.8.6), if V is a finitely generated projective T-module, the exterior algebra $\bigwedge_T(V)$ of V over T is strongly Gorenstein. If $V \neq 0$ it is not of finite global dimension. See also (8.5.7) for more details on this example.

---

[2] Confer (C.5.9).

[3] Affine.

[4] The "only if" direction does not hold unless S is commutative, for example T could be a field and S a strongly Gorenstein but not self injective algebra.

84 The second criterion has been mentioned already in the remark of (4.1). It is due to J.-E. Roos, [**Roo73**]:

PROPOSITION 8.3.3. *Let S be an arbitrary ring endowed with an ascending and exhaustive filtration $(F_i)_{i\geqslant 0}$ by subgroups such that*

- *$F_i \cdot F_j$ is contained in F_{i+j} for all i, j, and*
- *the associated graded ring* $\mathrm{gr}(S)$ *is strongly Gorenstein.*

Then S itself is strongly Gorenstein and its virtual dimension is bounded above by the virtual dimension of $\mathrm{gr}(S)$.

Furthermore, if M is any S-module allowing a filtration compatible with the given filtration on S and such that the associated $\mathrm{gr}(S)$*-module* $\mathrm{gr}(M)$ *is maximal Cohen–Macaulay, then M is already MCM over S.*

PROOF. The result follows essentially from the existence of a converging spectral sequence for any two filtered S-modules M and N for which the associated $\mathrm{gr}(S)$-module $\mathrm{gr}(M)$ is of finite type, starting with the E_1-terms

$$\mathrm{Ext}_{\mathrm{gr}(S)}(\mathrm{gr}(M), \mathrm{gr}(N))$$

and converging towards $\mathrm{Ext}_S(M, N)$, (see [**Roo73**] or [**Bjö79**, Ch.2, §3] for details). □

This result applies in particular to primitive quotients of enveloping algebras of finite dimensional Lie-algebras over a field in view of the Poincaré–Birkhoff–Witt theorem, cf. [**Roo73**] again. Unfortunately—for our purposes!—"most" of these rings seem to be already of finite global dimension, for example, this holds for the enveloping algebras themselves.

The first example of such a ring which is not of finite global dimension has been given by J. T. Stafford, [**Sta82**, Prop. 3.5]: $U(\mathrm{sl}_2\mathbb{C})/(C+1)$, where $C = H^2 + 2H + 4FE$ is the Casimir element, is of virtual dimension one, [**Roo73**, Cor. 2], but of infinite global dimension.

This example raises some interesting questions: As $C+1 = (H+1)^2+4FE$, the associated graded ring is the homogeneous coordinate ring of the quadric $z^2 + xy$, where z is the class of H (or $H+1$), x the class of $2E$ and y the class of $2F$ for example. Hence the associated graded ring has only an isolated singularity of type A_1 at the origin and it is well-known that this ring is of finite MCM-representation type. More precisely, see [**BEH87**], for this ring the category of (ungraded) MCMs modulo projectives is equivalent to the category of finite dimensional $\mathbb{C}$-vectorspaces.

Furthermore, it is not hard to see that the module of infinite projective dimen-
85 sion over the given primitive quotient of $U(\mathrm{sl}_2\mathbb{C})$ which was exhibited by J. T. Stafford in (loc. cit.) has as its associated graded module over the quadric precisely the unique indecomposable and non-free MCM over this A_1-singularity.

Hence, is it true that this module is the unique indecomposable and non-free MCM over this primitive quotient? Furthermore, given the correspondence between simple surface singularities over the complex numbers and semi-simple Lie-algebras as established by E. Brieskorn—see [**Slo80**]—and the fact that these are the only normal two-dimensional Gorenstein singularities which are of finite representation type—see [**BGS87**]—it would be interesting to know:

- Are there always primitive quotients of the enveloping algebras of the corresponding Lie-algebras whose category of MCMs modulo projectives is equivalent to the corresponding category over the associated simple singularity?
- Given the "explicit" knowledge of $\underline{\text{MCM}}$ for a simple singularity, see [**Knö87**], does an affirmative answer to the foregoing question yield a method to interpret these categories directly in terms of representations of the corresponding semi-simple Lie-algebra, adding another facet to the so-called "McKay-correspondence"?

Apart from this possible connection with simple singularities:

- Is it true in general that primitive quotients of enveloping algebras of (semi-simple) finite dimensional Lie-algebras over a field are always of finite MCM-representation type in the sense that only finitely many indistinct isomorphism classes of indecomposable MCMs exist for such rings? (This could then be interpreted as saying that such rings are never "too far" from being of finite global dimension.)

Much more and natural examples of strongly Gorenstein rings to which (8.3.3) applies and which are "almost never" of finite global dimension are obtained from *graded* or *super-Lie-algebras.*

Before treating these examples, we will shortly comment on the necessary—and rather obvious—modifications of the theory in

8.4 The graded case. As already mentioned several times and indicated in (4.8), most of the foregoing results on MCMs over rings which are strongly Gorenstein should hold in a much broader context. A simple extension, for which
all proofs remain literally the same, is furnished by considering *graded rings and* 86
modules.

Hence assume, in addition to the general hypotheses of being noetherian and of finite injective dimension on both sides, that $S = \bigoplus_{\delta\in\Delta} S_\delta$ is a Δ-graded ring for some commutative group of degrees $(\Delta, +)$. Then replace the various categories defined so far for ungraded rings correspondingly by $\mathrm{grMod}(S)$, $\underline{\mathrm{grmod}}(S)$, $\mathrm{D}^*_{\mathrm{gr}}(S)$, $\underline{\mathrm{D}}^{\mathrm{b}}_{\mathrm{gr}}(S)$, ..., constructed out of the category of Δ-graded S-modules with degree preserving S-linear maps as morphisms. Accordingly, one defines graded complete resolutions, graded perfect complexes and so on, to obtain mutatis mutandis the "same" results for such graded strongly Gorenstein rings, results we will henceforward take for granted.

As usual, to compensate for the lack of sufficiently many morphisms, one introduces shift functors $-(\delta)$ for all elements δ of Δ, which may depend on the choice of a commutation factor ε on $\Delta \times \Delta$, (see [**Bou70**, III.47ff]). Given then a graded S-module M, the grading of the shifted module is given by $M(\delta)_{\delta'} = M_{\delta+\delta'}$, and its (right) S-module structure is modified according to

$$m(\delta) \cdot s = m \cdot \varepsilon(\sigma, \delta) \cdot s \quad \text{in } M(\delta)_{\mu+\sigma} = M_{\delta+\mu+\sigma}$$

for homogeneous elements $m \in M_{\delta+\mu}$ and $s \in S_\sigma$.

These shift functors do not alter morphisms and preserve obviously graded perfect complexes, complete resolutions and MCMs.

As they also commute with the translation functor on complexes, one can consequently define Δ-*graded Tate cohomology* groups by

$$\underline{\mathrm{Ext}}^i_S(X, Y)_\delta = \mathrm{H}^i(\mathrm{Hom}_S(\mathbf{CR}(X), Y(\delta))) \tag{8.4.1}$$

for all integers i, degrees δ in Δ and complexes of graded S-modules Y in $\mathrm{D}^-(\mathrm{grMod}(S))$ and X in $\mathrm{D}^{\mathrm{b}}_{\mathrm{gr}}(S)$ as in (7.4.1).

The Yoneda products are compatible with the shift functors which may be expressed equivalently by saying that $\mathrm{D}^*(\mathrm{grMod}(S))$, $\mathrm{D}^{\mathrm{b}}_{\mathrm{gr}}(S)$, $\underline{\mathrm{D}}^{\mathrm{b}}_{\mathrm{gr}}(S)$ and its equivalent companions are in fact $\mathbb{Z}\times\Delta$-graded categories.

The multiplicative structure of these categories depends on the chosen commutation factor: The graded endomorphism ring of S as a graded right mod-
87 ule over itself, $\bigoplus_\delta \mathrm{Hom}_S(S, S(\delta))$, is not S again in general. One certainly has $\mathrm{Hom}_S(S, S(\delta)) = S_\delta$ as abelian groups, the identification obtained as usual by evaluating a homomorphism in the unit of the ring S, (which is tacitly assumed to be an element of S_0).

To an element s in S_δ corresponds then the homomorphism h_s given by

$$h_s(s') = h_s(1)\cdot s' = \varepsilon(\delta',\delta)ss' \quad \text{for } s' \text{ in } S_{\delta'}$$

as h_s is right S-linear and in view of the definition of the module structure on $S(\delta)$. It follows that in terms of this identification, the "Yoneda product" on S, considered as its own endomorphism ring, satisfies

$$s\circ s' = \varepsilon(\delta',\delta)ss' \quad \text{for } s \text{ in } S_\delta,\ s' \text{ in } S_{\delta'}.$$

Consequently, the graded S-dual of a graded S-module M, given by

$$M^* = \bigoplus_\delta \mathrm{Hom}_S(M, S(\delta))$$

as a Δ-graded abelian group, becomes a graded *left* module over this endomorphism ring—or, equivalently, it carries the structure of a graded *right* module over the graded ε-opposite ring ${}_\varepsilon S^{\mathrm{op}}$ of S, whose multiplication, say "#", is determined by

$$s'\#s := \varepsilon(\delta',\delta)ss' \quad \text{for } s \text{ in } S_\delta,\ s' \text{ in } S_{\delta'}.$$

Hence all statements concerning duality, like in (4.6), (6.2) or Chapter 7, have to be interpreted over this ε-opposite ring ${}_\varepsilon S^{\mathrm{op}}$. The details are left to the reader; as an example we only mention the extension of (7.1.2) to the graded case:

For any two classes $f \in \underline{\mathrm{Ext}}^i_S(Z,Y)_\delta$ and $g \in \underline{\mathrm{Ext}}^j_S(X,Z)_{\delta'}$, one finds

$$(8.4.2) \qquad (f\circ g)^* = (-1)^{ij}\varepsilon(\delta',\delta)g^*\circ f^* \quad \text{in } \underline{\mathrm{Ext}}^i_{{}_\varepsilon S^{\mathrm{op}}}(Y^*,X^*)_{\delta+\delta'}.$$

If we take $X = Y = Z$, and extend the commutation factor ε on Δ to the commutation factor ε' on $\mathbb{Z}\times\Delta$ by

$$\varepsilon'((i,\delta),(j,\delta')) := (-1)^{ij}\varepsilon(\delta,\delta')$$

for all integers i, j and degrees δ, δ' in Δ, the $\mathbb{Z}\times\Delta$-graded "stabilized Yoneda-Ext-algebra" $\underline{\mathrm{Ext}}_S(X,X)$ is seen to be isomorphic to the graded ε'-opposite algebra of $\underline{\mathrm{Ext}}_{{}_\varepsilon S^{\mathrm{op}}}(X,X)$.

88 After these brief comments on the necessary modifications for graded rings which are strongly Gorenstein, we can treat the example of

8.5 Graded or super-Lie-algebras. Let $(\Delta,+)$ be a commutative group (of degrees), $|\cdot|\colon \Delta \to \mathbb{Z}/2\mathbb{Z}$ a group homomorphism, Δ^+ its kernel, also called the group of *even* degrees, and Δ^- its complement, consisting of all *odd* degrees.

Given a commutative ring K, a Δ-graded super-K-Lie-algebra consists of a Δ-graded K-module $L = \bigoplus_{\delta\in\Delta} L_\delta$ endowed with bilinear pairings

$$[\,,]\colon L_\delta \times L_{\delta'} \longrightarrow L_{\delta+\delta'} \quad \text{for all pairs of degrees } (\delta,\delta'),$$

and quadratic maps

$$q\colon L_\delta \to L_{2\delta} \quad \text{for all odd degrees } \delta \text{ in } \Delta^-\,,$$

satisfying the usual axioms (considered by Milnor–Moore and amended by G. Sjödin; see [**Avr84**] for details—and replace $|a|$ there by $|\deg a|$ for any homogeneous element a in $L_{\deg a}$).

As always, examples of such Lie-algebras arise from Δ-graded, associative K-algebras A if one defines the Lie-bracket as

$$[a,b] := ab - (-1)^{|\delta||\delta'|}ba \quad \text{for all } a \in A_\delta \text{ and } b \in A_{\delta'}$$

and the quadratic maps as

$$q(a) := a^2 \quad \text{if } a \text{ is a homogeneous element of odd degree.}$$

The left adjoint to this "forgetful functor" associates then in the usual way to a Δ-graded super-K-Lie-algebra L its universal enveloping algebra $U(L)$, which is a Δ-graded, cocommutative K-Hopf-algebra with respect to the commutation factor $\varepsilon(\delta,\delta') = (-1)^{|\delta||\delta'|}$ on $\Delta \times \Delta$. It comes equipped with the natural "inversion" (or "antipodism"; [**CE99**]), which is the isomorphism of Δ-graded Hopf-algebras from $U(L)$ onto its ε-opposite (Hopf-)algebra ${}_\varepsilon U(L)^{\mathrm{op}}$ induced by the (ordinary) opposite of the identity on L. This enables one in particular to consider the graded dual of a (right) $U(L)$-module as a (right) $U(L)$-module again.

Finally, denoting by L^+ the (ordinary and Δ^+-graded) sub-Lie-algebra of L spanned by all homogeneous elements of even degree, and by L^- the K-submodule **89** spanned by those of odd degree, $L = L^+ \oplus L^-$ is the underlying $\mathbb{Z}/2\mathbb{Z}$-graded (or "super"-)K-Lie-algebra of L, and, accordingly, $U(L)$ the associated enveloping "super"-algebra.

If now L is finitely generated projective as a K-module, the canonical increasing filtration by tensor degree on $U(L)$ yields as the associated graded ring (or K-algebra) $\mathrm{Sym}(L^+) \otimes_K \bigwedge(L^-)$, the ($\mathbb{Z} \times \Delta$-graded) tensor product of the symmetric (K-)algebra on L^+ with the exterior (K-)algebra on L^-, by the graded version of the Poincaré–Birkhoff–Witt theorem. Remark that this associated graded algebra is nothing but the enveloping algebra of the underlying abelian Lie-algebra of L, as usual.

To apply (8.3.3), we only need the following

LEMMA 8.5.1. *Let K be any commutative ring with connected prime spectrum, V and W finitely generated projective K-modules of rank v and w respectively. Set $S = \mathrm{Sym}(V) \otimes_K \bigwedge(W)$. Then*

(i) *For the canonical augmentation module K of S, one has*

$$\mathrm{Ext}^i_S(K,S) = \begin{cases} 0 & \text{for } i \neq v \\ \det(W) \otimes_K \det(V') & \text{for } i = v\,. \end{cases}$$

as K-modules, where V' denotes the K-dual of V, $\det(-)$ the "determinant"—or highest non-vanishing exterior power—of a finitely generated projective K-module.

(ii) *If K is furthermore Gorenstein of finite Krull dimension k then S is strongly Gorenstein of virtual dimension $k + v$. It is of finite global dimension if and only if this holds for K and in addition $W = 0$.*

PROOF. (i) follows easily from the change-of-rings spectral sequence

$$\mathrm{Ext}^i_{\bigwedge(W)}(K, \mathrm{Ext}^j_S(\bigwedge(W), S)) \implies \mathrm{Ext}^{i+j}_S(K, S)$$

obtained from the obvious K-algebra homomorphisms $S \to \bigwedge(W) \to K$. The S-module $\bigwedge(W)$ is resolved by the Koszul complex over the S-linear map $V \otimes_K S \to S$ which identifies the generating set V with the subset $\mathrm{Sym}_1(V) \otimes_K 1$ of S. Consequently, $\mathrm{Ext}^j_S(\bigwedge(W), S) = 0$ for $j \neq v$, and $\mathrm{Ext}^v_S(\bigwedge(W), S) = \bigwedge(W) \otimes_K (\det V)'$. But $\mathrm{Ext}^i_{\bigwedge(W)}(K, \bigwedge(W)) = 0$ for $i \neq 0$ and $\mathrm{Hom}_{\bigwedge(W)}(K, \bigwedge(W)) = \det(W)$, which shows that the spectral sequence degenerates to yield (i).

90 For (ii), observe that $\mathrm{Sym}(V)$ is (locally) a polynomial ring over K which is Gorenstein if and only if K has this property. Furthermore, its Krull dimension is $\operatorname{rank} V + \dim K$. Now S itself is a finite flat $\mathrm{Sym}(V)$-algebra with respect to the natural inclusion of $\mathrm{Sym}(V)$ as the first factor of S. Hence one may conclude by (8.3.2), taking into account that the formation of exterior powers commutes with any change of rings and that over a field the exterior algebra of a finite dimensional vectorspace is quasi-Frobenius, and furthermore of infinite global dimension as soon as the vectorspace is non-zero. □

In view of this lemma, (8.3.3) yields

PROPOSITION 8.5.2. *Assume given a commutative Gorenstein ring K of finite Krull dimension k, $(\Delta, +)$ a group of degrees with a "parity homomorphism" $|\cdot|\colon \Delta \to \mathbb{Z}/2\mathbb{Z}$ as above, L a Δ-graded K-super-Lie-algebra such that the underlying K-modules of L^+ and L^- are both finitely generated projective and have a rank, say l^+ and l^- respectively. Then*

(i) *$U(L)$ as a Δ-graded ring is strongly Gorenstein of injective dimension equal to $k + l^+$. Its global dimension is bounded below by the global dimension of K.*
(ii) *The (total or derived) graded dual of the natural augmentation module K of $U(L)$ as an object in $\mathrm{D^b}({}_\varepsilon U(L)^{\mathrm{op}})$ satisfies*
$$K^* := \mathrm{RHom}_{U(L)}(K, U(L)) = \det(L^-) \otimes_K \det(L^+)'[-l^+]\,.$$
(iii) *If K is regular, hence of finite global dimension, for $U(L)$ to be of infinite global dimension, it is necessary that $l^- \neq 0$ and it is sufficient that L^- has non-zero intersection with the center of L.*

PROOF. Applying (8.3.3) to the filtration by tensor degree on $U(L)$, the lemma above yields a priori that $U(L)$ is strongly Gorenstein of virtual dimension at most equal to $k + l^+$. That one has indeed equality follows from (ii) which in turn is a consequence of the (indicated) proof of (8.3.3) and (8.5.1) (i): Taking $M = K$ and $N = U(L)$, the associated graded module of M is K again and that of N is the associated graded ring of $U(L)$. But then (8.5.1) (i) implies that the spectral sequence in the proof of (8.3.3) for these $U(L)$-modules degenerates, leaving

$$\mathrm{Ext}^{l^+}_{U(L)}(K, U(L)) = \mathrm{Ext}^{l^+}_{\mathrm{gr}(U(L))}(K, \mathrm{gr}(U(L))) = \det(L^-) \otimes_K \det(L^+)'$$

91 as the only possibly non-vanishing cohomology group of K^*, which is precisely the assertion in (ii). The last statement in (i) follows immediately from the fact that $U(L)$ is a supplemented flat K-algebra: Whenever A, B are two K-modules, one has

$$\mathrm{Ext}_K(A, B) = \mathrm{Ext}_{U(L)}(A \otimes_K U(L), B)\,.$$

B considered a (right) $U(L)$-module via the augmentation.

The first part of (iii) follows from the last statement of (8.5.1) as the global dimension of $U(L)$ is also bounded above by the global dimension of its associated graded ring, once again using J.-E. Roos' result as in (8.3.3); cf. also [**Bjö79**, Ch.2, Thm.3.7.].

The proof of the remaining assertion of (iii) is obtained as follows: Without loss of generality[5], one may assume that K is a field and that there exists a homogeneous element u in L^- which is non-zero and in the center of L. If then $\underline{L}$ denotes the still Δ-graded K-super-Lie-algebra which is the quotient of L by the ideal Ku, the universal enveloping algebra $U(L)$ is isomorphic to $U(\underline{L}) \otimes_K \bigwedge(Ku)$, and the $U(L)$-module $U(\underline{L})$ is easily seen to be of infinite projective dimension, a projective resolution being obtained by tensoring a resolution of K over $\bigwedge(Ku)$ with $U(L)$ over K. In other words, one uses the well-known relation: $\operatorname{Ext}_{\bigwedge(Ku)}(K,K) = \operatorname{Ext}_{U(L)}(U(\underline{L}),K)$, (see [**CE99**, XIII.4.4.] for example). □

Now let us set by (slight) analogy with the case of integral group representations

$$\widehat{\mathrm{H}}^i(L/K,M) := \underline{\operatorname{Ext}}^i_{U(L)}(K,M) \tag{8.5.3}$$

for any (complex of) right, graded $U(L)$-module(s) M in $\mathrm{D}^-(\operatorname{Mod}(U(L)))$ and any integer i, and call the groups the *Tate cohomology* of L with values in M. These groups are naturally Δ-graded K-modules and the Yoneda product defines on the total Tate cohomology $\widehat{\mathrm{H}}(L/K,M)$ the structure of a right $\mathbb{Z}\times\Delta$-graded module over the *Tate cohomology ring* of L, $\widehat{\mathrm{H}}_L := \widehat{\mathrm{H}}(L/K,K)$.

Recall furthermore that $\mathrm{H}^*(L/K,M) := \operatorname{Ext}_{U(L)}(K,M)$ is the ordinary cohomology of L with values in M, and that the homology of L with values in M is given by $\mathrm{H}_*(L/K,M) := \operatorname{Tor}^{U(L)}(M,K)$.

Finally, set $\det(V^+\oplus V^-) := \det(V^+)\otimes_K\det(V^-)'$, the "*super-determinant*"—or "*Berezinian*" in super-terminology—for any $\mathbb{Z}/2\mathbb{Z}$-graded K-module, where V^+, **92**
the submodule of even elements, and V^-, the submodule of odd elements, are both finitely generated projective. Remark that the "super-determinant" is a projective K-module of rank one over K and that its inverse (in the Picard group of K) is represented by its K-dual or, as well, as both the "super-determinant" of the graded module $V^-\oplus V^+$, which is obtained from the original one by "parity-change", or of the graded K-dual $(V^+)'\oplus(V^-)'$.

Now, combining (8.5.2) (ii) and (6.2.5) 3, we get for any object M in $\mathrm{D}^-(\operatorname{Mod}(U(L)))$ a long exact sequence of Δ-graded K-modules

$$\begin{aligned} \cdots \longrightarrow \widehat{\mathrm{H}}^{i-1}(L/K,M) \longrightarrow \mathrm{H}_{-i+l^+}(L/K,M)\otimes_K \det(L)^{-1} \longrightarrow \\ \xrightarrow{\mathbf{N}^i} \mathrm{H}^i(L/K,M) \xrightarrow{c^i} \widehat{\mathrm{H}}^i(L/K,M) \longrightarrow \cdots, \end{aligned} \tag{8.5.4}$$

where $\mathbf{N}^i$ is the morphism induced by the Norm map, c^i the natural transformation from ordinary to stabilized or Tate cohomology.

In particular, one has—as in (6.3.6)—for any graded $U(L)$-module

$$\widehat{\mathrm{H}}^i(L/K,M) = \begin{cases} \mathrm{H}^i(L/K,M) & \text{for } i > l^+ \\ \mathrm{H}_{l^+-i}(L/K,M)\otimes_K \det(L)^{-1} & \text{for } i < -1\,. \end{cases}$$

[5]For example, by localizing at a minimal prime.

Just remark that

$$\begin{aligned}
\operatorname{Tor}_{-i}^{U(L)}(M, K^*) &= \mathrm{H}^i(M \otimes^{\mathsf{L}}_{U(L)} \mathrm{RHom}_{U(L)}(K, U(L))) && \text{by definition,}\\
&= \mathrm{H}^i(M \otimes^{\mathsf{L}}_{U(L)} \det(L)^{-1}[-l^+]) && \text{by (8.5.2) (ii),}\\
&= \mathrm{H}^{i-l^+}(M \otimes^{\mathsf{L}}_{U(L)} K) \otimes_K \det(L)^{-1} && \text{as } \det(L)^{-1} \text{ is projective over } K,\\
&= \operatorname{Tor}_{-i+l^+}^{U(L)}(M, K) \otimes_K \det(L)^{-1} && \text{again by definition.}
\end{aligned}$$

Next, we want to consider the duality theory of Chapter 7 in this situation. To avoid (notational) complications arising from the structure of K, assume henceforth that K is (commutative) semi-simple—hence a product of fields. Then, given any object M in $\mathrm{D}^-(\mathrm{Mod}(U(L)))$, let $M^\sharp$ denote the graded K-dual of M, converted into a complex of right $U(L)$-modules again by use of the inversion.

93 Now the graded version of Theorem 7.5.1 or its Corollary 7.5.2 reads:

PROPOSITION 8.5.5. *For any object N in $\mathrm{D}^{\mathrm{b}}(U(L))$ there are natural isomorphisms for any integer i*

$$\widehat{\mathrm{H}}^{i+l^+-1}(L/K, N^\sharp) \otimes_K \det(L) \longrightarrow (\widehat{\mathrm{H}}^{-i}(L/K, N))'.$$

PROOF. Apply (7.5.3) with $T = K$, $S = {}_\varepsilon U(L)^{\mathrm{op}}$ and $W = K$ to get isomorphisms

$$\underline{\mathrm{Ext}}^{i-1}_{{}_\varepsilon U(L)^{\mathrm{op}}}(K^*, \mathrm{Hom}_K(N, K)) \longrightarrow \mathrm{Hom}_K(\underline{\mathrm{Ext}}^{-i}_{U(L)}(K, N), K)$$

as K is injective as a module over itself. Using then once again the known structure of K^*, (8.5.2) (ii), as above, and converting the occurring graded ${}_\varepsilon U(L)^{\mathrm{op}}$-modules into $U(L)$-modules by means of the inversion exhibits the terms in the form claimed. □

Finally, let us examine a little bit closer $\widehat{\mathrm{H}}_L$, the Tate cohomology ring of a graded super-Lie-algebra L, over a field.

PROPOSITION 8.5.6. *Let K be a field, L a Δ-graded super-Lie-algebra over K, $|\cdot|\colon \Delta \to \mathbb{Z}/2\mathbb{Z}$ a parity homomorphism as before.*

(i) *The $\mathbb{Z} \times \Delta$-graded Tate cohomology ring $\widehat{\mathrm{H}}_L$ of L is ε'-commutative, where ε' is the commutation factor*

$$\varepsilon'((i,\delta),(j,\delta')) = (-1)^{ij}\varepsilon(\delta,\delta') = (-1)^{ij+|\delta||\delta'|}$$

for pairs (i,δ) and (j,δ') in $\mathbb{Z} \times \Delta$. In particular $\widehat{\mathrm{H}}^0_L = \underline{\mathrm{Hom}}_{U(L)}(K, K)$ is a Δ-graded, ε-commutative K-algebra.

(ii) *$\widehat{\mathrm{H}}_L$ is a self-injective K-algebra. More precisely, there is a non-degenerate K-bilinear pairing, associative for the Yoneda product on $\widehat{\mathrm{H}}_L$, and homogeneous for the total Δ-degree:*

$$\widehat{\mathrm{H}}_L^{*+l^+-1} \otimes_K \widehat{\mathrm{H}}_L^{-*} \longrightarrow \det(L)^{-1}.$$

PROOF. As seen in (8.4.2),

$$\underline{\mathrm{Ext}}_{U(L)}(K, K) = \widehat{\mathrm{H}}_L$$

is ε'-opposite to $\underline{\mathrm{Ext}}_{{}_\varepsilon U(L)^{\mathrm{op}}}(K^*, K^*)$. Using the inversion on $U(L)$ and the fact that
94 $\mathrm{RHom}_{U(L)}(K, K)$ and $\mathrm{RHom}_{U(L)}(K^*, K^*)$ are canonically isomorphic, one gets an

ε'-anti-isomorphism of $\widehat{\mathrm{H}}_L$ onto itself. The point now is, that this map is in fact the identity, implying the claimed ε'-commutativity. The details are left to the reader. They can easily be filled in, as there is the so-called "standard-resolution" of K over $U(L)$—see below or [**FHJ$^+$88**]—to which the foregoing functors and identifications can be applied directly.

Here, let us only remark that (i) "stabilizes" the classical result which states that the (ordinary) cohomology ring H_L itself is ε'-commutative; see for example [**BR86**, §0] and the sources cited there.

The proof of (ii) is obvious from (8.5.5) above: Just take $N = K$—so that $N^\sharp$ equals K again—to obtain isomorphisms of K-modules

$$\widehat{\mathrm{H}}_L^{i+l^+-1} \otimes_K \det(L) \xrightarrow{\cong} (\widehat{\mathrm{H}}_L^{-i})'$$

for all i. Such an isomorphism corresponds biuniquely to a non-degenerate K-bilinear pairing as claimed. That these pairings are associative with respect to the (Yoneda) product on the total Tate cohomology ring, follows from (7.5.4)—as in (7.7.5) (iii)—in its graded version; cf. (8.4.2). □

REMARKS. (a) Assume chosen Δ-homogeneous K-bases $\{y_1, \ldots, y_{l^-}\}$ of L^- and $\{x_1, \ldots, x_{l^+}\}$ of L^+ respectively. Then the one-dimensional vectorspace $\det(L)$ is generated formally by the "super-volume element"

$$\mathrm{vol}(L) := \frac{x_1 \wedge x_2 \wedge \cdots \wedge x_{l^+}}{y_1 \wedge y_2 \wedge \cdots \wedge y_{l^-}}.$$

In other words, $\det(L)^{-1}$ is isomorphic as a Δ-graded K-vectorspace to

$$K(\mathrm{vol}(L)^{-1})K\left(\sum_j \deg y_j - \sum_i \deg x_i\right),$$

where $\deg(\cdot)$ denotes the degree in Δ of a homogeneous element.

(b) The pairing in (ii) above stabilizes the usual pairing between ordinary cohomology and homology of L: The following diagram is commutative, the horizontal arrows being isomorphisms of graded K-vectorspaces

$$\begin{array}{ccc}
\widehat{\mathrm{H}}_L^{i+l^+-1} \otimes_K \det(L) & \xrightarrow{\cong} & (\widehat{\mathrm{H}}_L^{-i})' \\
\downarrow & & \downarrow{\scriptstyle (c^{-i})'} \\
\mathrm{H}_{-i}(L/K, K) & \xrightarrow{\cong} & (\mathrm{H}_L^{-i})'
\end{array}$$

the left vertical map being the connecting homomorphism of (8.5.4) for $M = K$, **95**
tensored with the identity on $\det(L)$.

The main source—in algebra and topology—of such graded super-Lie-algebras is provided by taking $\Delta = \mathbb{Z}$ and—see [**Avr84**], [**Qui69**] or [**FHJ$^+$88**] for details—considering either

- the *homotopy-Lie-algebra* $\pi^*(R)$ of an augmented commutative K-algebra $R \to K$, or
- the *rational homotopy-Lie-algebra* $\pi_*(\Omega X) \otimes \mathbb{Q}$ of a simply connected topological space X with the Samelson product as bracket.

In the first case, the enveloping algebra is by definition $\mathrm{Ext}_R(K, K)$, the usual Yoneda-Ext-algebra of R, and, if R happens to be a graded "Koszul-algebra", [**Löf86**], R itself with its grading is the cohomology ring of its homotopy-Lie-algebra.

In the second case, the enveloping algebra is the rational homology of the loop-space ΩX of X. Here, it would be interesting to know, what the topological or homotopy-theoretic interpretation of MCMs and Tate cohomology might be.

These examples show once again—similar to ((8.2.2)) & (8.2.3)—a shortcoming of the theory here: Most super-Lie-algebras arising these ways are not finite dimensional but rather locally finite in the sense that each graded piece has this property. Given the results of [**FHJ$^+$88**], where it is shown that the injective dimension of the corresponding enveloping algebra can be bounded above by the L.S.-category of the topological space, at least those where the L.S.-category is finite should be enclosed into a more general theory.[6]

We conclude with the simple

EXAMPLE 8.5.7 (Abelian super-Lie-algebras). We return to the notations of (8.5.1): L is the abelian super-Lie-algebra over a field K with a v-dimensional vectorspace V in even degrees and a w-dimensional vectorspace W in odd degrees, the enveloping algebra $\mathrm{Sym}(V) \otimes_K \bigwedge(W)$ denoted S again.

A complete projective resolution of K over S is obtained—as in (5.6)—by connecting a projective resolution of K with the ${}_\varepsilon S^{\mathrm{op}}$-dual of a projective resolution of $K^* = (\det(W) \otimes_K \det(V'))[-v]$, (8.5.2) (ii).

96 The natural projective resolution of K over S is well-known: First, if $W = 0$, K is resolved projectively over $\mathrm{Sym}(V)$ by the ordinary Koszul complex

$$\mathbf{P}_{\mathrm{Sym}(V)}(K) := (\bigwedge(dV) \otimes_K \mathrm{Sym}(V), d_V)\,,$$

where dV denotes an isomorphic copy of V, with the same Δ-grading but placed in complex-degree -1 (according to our general convention on complexes). In other words, dV is a $\mathbb{Z} \times \Delta$-graded K-vectorspace, concentrated in degrees $\{-1\} \times \Delta$. "Extending" the original Δ-grading on V to a $\mathbb{Z}\times\Delta$-grading with V concentrated in degrees $\{0\}\times\Delta$, the underlying $\mathbb{Z}\times\Delta$-graded K-vectorspace of the Koszul complex becomes an ε'-commutative K-algebra with respect to the commutation factor ε' given in (8.5.6).

Then, as is well-known, [**Bou80**, 206; Exerc. (2)], d_V is the unique K-algebra derivation of bi-degree $(1, 0)$ extending the K-linear endomorphism of $dV \oplus V$ which maps dV isomorphically onto V and annihilates V.

As $\mathrm{RHom}_{\mathrm{Sym}(V)}(K, \mathrm{Sym}(V)) = \det(V)^{-1}[-v]$, the algorithm of (5.6) yields the natural projective co-resolution

$$\mathbf{C}_{\mathrm{Sym}(V)}(K) = (\det(V) \otimes_K \bigwedge(dV') \otimes_K \mathrm{Sym}(V)[v], d_V^\sharp)\,,$$

where $(dV)'$, the graded K-dual of dV is accordingly concentrated in degrees $\{1\} \times \Delta$, and the differential $d_V^\sharp$ is the $\mathrm{Sym}(V)$-dual of d_V, shifted v times and multiplied with the identity on $\det(V)$.

Remark that as $\mathbb{Z} \times \Delta$-graded vectorspaces one has

$$\det(V)[v] \cong \det(dV)$$

[6]This is incorrect—[**FHJ$^+$88**] shows the depth is bounded above by the L.S.-category, not by the injective dimension, which may be infinite.

and a natural isomorphism ([**Bou80**, 149] for example),

$$\bigwedge(dV) \longrightarrow \det(dV) \otimes_K \bigwedge(dV')$$

as usual.

This isomorphism on the generating sets extends naturally to an isomorphism of complexes from $\mathbf{P}_{\mathrm{Sym}(V)}(K)$ onto $\mathbf{C}_{\mathrm{Sym}(V)}(K)$ and represents a natural Norm map $\mathbf{N}_{\mathrm{Sym}(V)}(K)$ in this case. Of course, the deduced complete resolution, (5.6.2), is contractible in accordance with the fact that K is of finite projective dimension over the polynomial ring $\mathrm{Sym}(V)$.

These classical properties of the Koszul complex have a (less known) "shifted" counterpart, first noted[7] by D. Quillen in this context, see [**Ill71**, I.4.3] for a com- **97**
plete treatment:

Namely, if now $V = 0$, so that we start with an abelian super-Lie-algebra concentrated in odd degrees, the enveloping algebra is $\bigwedge(W)$, (which is in particular quasi-Frobenius), and the natural projective resolution of K over it is given by the "shifted" Koszul complex

$$\mathbf{P}_{\bigwedge(W)}(K) := (\Gamma(dW) \otimes_K \bigwedge(W), d_W)\,,$$

dW denoting a copy of W concentrated in degrees $\{-1\} \times \Delta$ as above, $\Gamma(dW)$ the K-algebra of divided powers over dW. Remark that in this situation, $\Gamma(W)$ is (by definition) isomorphic to the graded K-dual of $\mathrm{Sym}(W')$—even as a graded Hopf-algebra.

Now the $\mathbb{Z} \times \Delta$-graded K-vectorspace underlying $\mathbf{P}_{\bigwedge(W)}(K)$ carries also a natural ε'-commutative K-algebra structure with respect to which d_W is the unique K-linear derivation extending the obvious K-linear map as before.

As here $\mathrm{RHom}_{\bigwedge(W)}(K, \bigwedge(W)) = \det(W)[0]$, the algorithm of (5.6)—see also (8.1)—shows that the natural projective co-resolution of K is

$$\mathbf{C}_{\bigwedge(W)}(K) := (\mathrm{Sym}(dW') \otimes_K \det(W)^{-1} \otimes_K \bigwedge(W), d_W^{\sharp})\,,$$

which is also a (minimal) injective resolution of K embedded by the map $K \xrightarrow{\cong} \det(W)^{-1} \otimes_K \det(W)$ into $\det(W)^{-1} \otimes_K \bigwedge(W)$ as its socle.

In this case of course, the Norm map $\mathbf{N}_{\bigwedge(W)}(K)$ is just the composition of the $\bigwedge(W)$-linear maps

$$\mathbf{P}_{\bigwedge(W)}(K) \xrightarrow{\text{augmentation}} K \xrightarrow{\cong} \det(W)^{-1} \otimes_K \det(W) \xrightarrow{\text{incl.}} \mathbf{C}_{\bigwedge(W)}(K)\,.$$

Having hence clarified the situation in the "pure" case, where the super-Lie-algebra is concentrated in one parity and abelian, one obtains in general:

For an abelian super-Lie-algebra $L = V \oplus W$, the projective (co-)resolution is obtained as the tensor product over K of the corresponding objects for V and W alone, considered as subalgebras of L. The associated Norm map $\mathbf{N}_S(K)$ is the tensor product of the corresponding Norm maps $\mathbf{N}_{\mathrm{Sym}(V)}(K)$ and $\mathbf{N}_{\bigwedge(W)(K)}$.

It follows then easily: The Tate cohomology ring of $L = V \oplus W$ is given by **98**

(i) $\widehat{\mathrm{H}}_L = 0$ iff $W = 0$, as precisely in this case S is of finite global dimension.
(ii) If $W \neq 0$, the exact sequence (8.5.4) for $M = K$ breaks up into short exact sequences as the corresponding (cohomological) Norm maps $\mathbf{N}^i$ are zero. Accordingly, as $\mathbb{Z} \times \Delta$-graded K-vectorspaces, one has that Tate cohomology is

[7] This is in fact the Cartan resolution of the exterior algebra; see [**43**, Vol. 3].

the direct sum of the ordinary cohomology and the shifted (by $[-v+1]$) and twisted (by $\det(L)^{-1}$) ordinary homology of L.

In terms of the explicit descriptions just mentioned, the multiplicative structure is as follows:

(iii) If $\dim_K W = 1$, say $W = K(\delta)$, the shifted vectorspace dW is generated by a $\mathbb{Z} \times \Delta$-homogeneous element of bi-degree $(-1, -\delta)$, say t^{-1}. Then, as a $\mathbb{Z} \times \Delta$-graded K-algebra, one finds

$$\widehat{\mathrm{H}}_L = \bigwedge(dV')[t, t^{-1}],$$

the algebra of Laurent-polynomials in t over the exterior K-algebra generated by dV'. In this identification, the pairing of (8.5.6) (ii) is given by the "super-residue", which associates to a pair of such polynomials the coefficient of $\mathrm{vol}(dV') \cdot t^{-1}$ in their product.

(iv) If $\dim_K W > 1$, the Tate cohomology ring is indeed the trivial algebra extension of the ordinary cohomology by the (shifted and twisted) ordinary homology considered as a bimodule. Explicitly:

$$\widehat{\mathrm{H}}_L = \bigwedge(dV') \otimes_K (\mathrm{Sym}(dW') \oplus \det(W) \otimes_K \Gamma(dW)[1]),$$

the graded tensor product of the exterior K-algebra on dV' with the trivial ring-extension of the symmetric algebra $\mathrm{Sym}(dW')$ by the indicated bimodule, where $\mathrm{Sym}(dW')$ acts on its graded K-dual $\Gamma(dW)$ by "contraction" or "inner products", as defined for example in [**Bou70**, III, §11, no. 6]. The pairing (8.5.6) (ii) is then again given by multiplication and then projection onto the super-volume element chosen. (Remark that $\det(L)^{-1}$ occurs as the direct summand $\bigwedge^v(dV') \otimes_K \det(W)$ in $\widehat{\mathrm{H}}_L^{v-1}$.)[8]

(v) The foregoing description of the pairings is in accordance with (7.7.5): Taking $T = \mathrm{Sym}(V)$, and $\omega_T = \Omega^v_{T/K} = \det(V) \otimes_K \mathrm{Sym}(V)$, the module of (Kähler-) differential forms of maximal degree of T over K, as dualizing module, and $f \colon T \to \mathrm{Sym}(V) \otimes_K \bigwedge(W)$ the obvious inclusion, then the associated dualizing
99 module of S is given by Remark (b) in (7.6):

$$\begin{aligned}
\omega_S &= f^! \omega_T[0] \\
&= \mathrm{Hom}_{\mathrm{Sym}(V)}(\mathrm{Sym}(V) \otimes_K \textstyle\bigwedge(W), \det(V) \otimes_K \mathrm{Sym}(V)) \\
&= (\mathrm{Sym}(V) \otimes_K \textstyle\bigwedge(W)) \otimes_K \det(W)^{-1} \otimes_K \det(V) \\
&= S \otimes_K \det(L)
\end{aligned}$$

as an S-bimodule, taking into account as before the natural identification of the graded K-dual of $\bigwedge(W)$ with $\bigwedge(W) \otimes_K \det(W)^{-1}$. Then the trace map of (7.7.5) (iv) for (the MCM-approximation of) K is the linear form

$$\tau_K \colon \underline{\mathrm{Ext}}_S^{v-1}({}^*K, \omega_S \otimes_S K^*) = \det(L) \otimes_K \widehat{\mathrm{H}}_L^{v-1} \longrightarrow K.$$

Observe that ${}^*K = K^*$ as complexes of S-bimodules and that $\omega_S \otimes_S K = \det(L)$ by the above—which exhibits the coefficient of $\det(L)^{-1}$ as a direct summand of $\widehat{\mathrm{H}}_L^{v-1}$.

[8] This anticipates [**18**]; see (C.5.9).

In "super-geometric" terms, this linear form is nothing but the integration against the "fundamental cycle" of the underlying projective super-space of S, as will become clear below.

This description of S as a finite flat $\mathrm{Sym}(V)$-algebra yields the following representation of MCMs over S: The polynomial algebra is concentrated in the even degrees Δ^+, so for any Δ-graded S-module M, the underlying $\mathrm{Sym}(V)$-module f_*M decomposes into a direct sum $f_*M = M^+ \oplus M^-$ of Δ^+-graded $\mathrm{Sym}(V)$-modules. By (8.3.1) in its graded version, the original module M is MCM over S iff both summand $M^{+/-}$ are finitely generated free $\mathrm{Sym}(V)$-modules.

The missing piece of information to reconstruct M from these free modules is the action of the 1-forms in $\bigwedge^1(W) = W$ as $\bigwedge(W)$ is freely generated as an alternating K-algebra by them.

Hence, an MCM over S can be represented as a collection

$$(M^+, M^-, A_i\colon M^+ \longrightarrow M^-, B_i\colon M^- \longrightarrow M^+; i = 1, \dots w),$$

where $M^{+/-}$ are graded free $\mathrm{Sym}(V)$-modules of finite rank, and A_i and B_i are $\mathrm{Sym}(V)$-linear maps satisfying the relations

$$A_iB_i = 0 = B_iA_i \quad \text{and} \quad A_iB_j + A_jB_i = 0 = B_iA_j + B_jA_i$$

for all $1 \le i, j \le w; i \ne j$.

We finish the section with two more **100**

REMARKS. (c) The given multiplicative structure of $\widehat{\mathrm{H}}_L$ implies that $\widehat{\mathrm{H}}_L^0$ is a *local* ring if $W \neq 0$. This proves that any MCM-approximation of K without free summands is indecomposable, (2.1), as an S-module as $\widehat{\mathrm{H}}_L^0$ is its endomorphism ring in $\underline{\mathrm{MCM}}(S)$. Furthermore, it has a well defined rank if constructed from the natural complete resolution, cf. (5.5.1). One may hence ask as in (5.5.3) whether this is the minimum for a non-free MCM-approximation of an S-module of finite length.

Also, it should be interesting to understand the representations of $\mathrm{Gl}(V) \times \mathrm{Gl}(W)$—or rather of the corresponding super-group—which occur in this indecomposable MCM-approximation of K, as this module determines essentially the Tate cohomology by (6.1.2), (6.4.1).

(d) For an arbitrary K-super-Lie-algebra L, extending naturally the Poincaré–Birkhoff–Witt theorem, one may obtain a standard (complete) resolution of K over $U(L)$ by deforming the projective (co-)resolution (and Norm map) of the underlying abelian super-Lie-algebra, see for example [**FHJ$^+$88**, Prop.1.13.] or [**Qui69**, App.B,6.7.] for the case of projective resolutions.

In other words, the underlying graded K-vectorspaces can be chosen to be the same, but the multiplication and differential become deformed.

It follows that, denoting L^{ab} the underlying abelian algebra, the Tate cohomology of L itself can be obtained as the cohomology of a differential, $\mathbb{Z} \times \Delta$-graded algebra (H, D), D a homogeneous differential of bi-degree $(1, 0)$, the underlying graded K-vectorspace of H being isomorphic to $\widehat{\mathrm{H}}_{L^{\mathrm{ab}}}$. In particular, the dimensions of the homogeneous components of $\widehat{\mathrm{H}}_L$ are bounded above by the corresponding dimensions in the abelian case, which can be easily determined from the explicit description above.

Although we ignore the meaning of MCMs for an arbitrary super-Lie-algebra, at least for certain of those, they admit a geometric interpretation. Here, we restrict ourselves again to the case of abelian super-Lie-algebras, referring the reader to [**BEH87**] for some more examples of the same kind. We want to show that over the category of abelian super-Lie-algebras—or, equivalently, graded K-vectorspaces—the theory of MCMs and of coherent sheaves on the associated projective superspaces are dual to each other. This extends naturally—and in a straightforward manner—the theory initiated by A. A. Beĭlinson, J. Bernšteĭn, S. I. Gelfand and I. M. Gelfand.

CHAPTER 9

Connection to geometry on projective super-spaces

Fix a field K and the group of degrees $\Delta = \mathbb{Z} \times \mathbb{Z}$ with parity homomorphism $|(a,b)| = b \bmod 2$ into $\mathbb{Z}/2\mathbb{Z}$. The total degree of (a,b) is given by $\int(a,b) = a + b$. **101**

9.1 Linear super-spaces. Consider the abelian category $\operatorname{mod}(K \times K)$ and interpret it as follows: Its objects, which are by definition pairs (V, W) of finite dimensional K-vectorspaces, are thought of as Δ-graded K-vectorspaces concentrated in total degree 1, the *even subspace* V being placed in degree $(1,0)$, the *odd subspace* W in degree $(0,1)$. For short, we will also say that (V, W) endowed with this grading is a linear super-space.

The morphisms $F = (F^+, F^-)\colon (V_1, W_1) \to (V_2, W_2)$ are accordingly just pairs of K-linear maps and can be interpreted as Δ-homogeneous morphisms of Δ-graded vectorspaces.

In particular, there are the following:

(i) A linear form λ on the even subspace V will be identified with the morphism $(\lambda, 0)\colon (V, W) \to (K, 0)$ and, if $\lambda \neq 0$, its kernel is said to define a (K-rational) *geometric point of a linear super-space* of (V, W).

(ii) An element w in the odd subspace W gives rise to a morphism $\langle w\rangle\colon (0, K) \to (V, W)$, $\langle w\rangle(0, a) = (0, wa)$, and its image, if not zero, will represent a (K-rational) *odd point of a linear super-space* of (V, W).

"Recall" that $\mathbb{P}(V) = \operatorname{Proj}_K(\operatorname{Sym}(V))$ is the scheme of the projective space of hyperplanes in V, so that "geometric points" of (V, W) correspond to the points of $\mathbb{P}(V)$, whereas the "odd points" of (V, W) are parametrized by $\mathbb{P}(W^*)$, $(-)^*$ denoting in this chapter the K-dual of a vectorspace.

(iii) A morphism F as above will be called (linearly) *perfect* iff its even component F^+ is surjective, whereas its odd component F^- is injective.

Two particular such morphisms are given by $i_V = (\mathrm{id}_V, 0)$, the inclusion of the even subspace $V = (V, 0)$ into (V, W), and by $p_W\colon (V, W) \to (0, W) = W$, the projection into the odd quotient-space.

Extending the notion of super-determinant for a $\mathbb{Z}/2\mathbb{Z}$-graded K-vectorspace, (8.5), define for any morphism F as above **102**

$$\det(F) := \det(V_2, W_2) \otimes_K \det(V_1, W_1)^{-1} = \det(\operatorname{Cok}(F)) \otimes_K \det(\operatorname{Ker}(F))^{-1}$$

and call it accordingly the *super-determinant* of F.

If $v_i = \dim_K(V_i)$ and $w_i = \dim_K(W_i)$, then, as graded K-vectorspaces,

$$\det(F) \cong K(v_1 - v_2, w_2 - w_1),$$

This chapter was originally titled: *Maximal Cohen–Macaulay Modules and Geometry on Projective Super-spaces (The Theory of Beĭlinson–Bernšteĭn–Gelfand–Gelfand)*

hence, if F is linearly perfect, $\det(F)$ is concentrated in negative degrees. We will also say that

$$\operatorname{ind}^+(F) := \operatorname{ind}(F^+) = \dim_K(\operatorname{Ker}(F^+)) - \dim_K(\operatorname{Cok}(F^+)) = v_1 - v_2$$

is the *even index* of F, whereas

$$\operatorname{ind}^-(F) := -\operatorname{ind}(F^-) = w_2 - w_1$$

represents the *odd index* of F.

On the category of linear super-spaces just defined, we consider the following *duality functor* $\mathbf{D}$, given by

$$\mathbf{D}(V,W) := (W^*, V^*) \quad \text{and} \quad \mathbf{D}(F^+, F^-) := ((F^-)^*, (F^+)^*) .$$

It is obviously a contravariant, exact functor satisfying $\mathbf{D}^2 = \mathrm{id}$, hence establishes indeed a "duality". It interchanges "geometric" and "odd" points and preserves linearly perfect morphisms. One has also $\operatorname{ind}^{+/-}(F) = \operatorname{ind}^{-/+}(\mathbf{D}(F))$.

9.2 The enveloping algebras and their modules. In accordance with the theory of "super-Lie-algebras" mentioned in the foregoing chapter, we have a functor $\mathbf{U}$ from the category $\operatorname{mod}(K \times K)$, alias linear super-spaces of abelian super-Lie-algebras, into the category of Δ-graded K-(Hopf-)algebras with inversion, given by

$$\mathbf{U}(V,W) := \operatorname{Sym}(V) \otimes_K \bigwedge(W) \quad \text{and} \quad \mathbf{U}(F^+,F^-) := \operatorname{Sym}(F^+) \otimes_K \bigwedge(F^-) .$$

103 These (Hopf-)algebras are ε-commutative (and ε-cocommutative) with respect to the commutation factor

$$\varepsilon((a_1,b_1),(a_2,b_2)) = (-1)^{b_1 b_2} .$$

Accordingly, we will not distinguish left and right $\mathbf{U}(V,W)$-modules, but rather assume that all modules are ε-symmetric bimodules, the left and right structure being obtained from one another by means of the inversion (or "graded commutativity").

By definition, $\mathbf{U}(V,W)$ is both a supplemented $\operatorname{Sym}(V)$-algebra as well as a supplemented $\bigwedge(W)$-algebra. In particular, $\operatorname{Sym}(V)$ and $\bigwedge(W)$ will always be considered $\mathbf{U}(V,W)$-modules with respect to the associated augmentations. Analogously, K just represents the canonical augmentation module of $\mathbf{U}(V,W)$.

Remark that the algebra homomorphism $\mathbf{U}(i_V)\colon \operatorname{Sym}(V) \to \mathbf{U}(V,W)$ exhibits $\mathbf{U}(V,W)$ as a finite flat $\operatorname{Sym}(V)$-algebra and that it is hence a duality morphism of finite type as defined in (7.6.2). Extending this notion and the result (7.6.3) to the "graded commutative" case—which is left to the reader—the term "linearly perfect morphism" is justified by

LEMMA 9.2.1. *Let $F = (F^+, F^-)$ be a pair of K-linear maps as above. Then the following properties are equivalent:*

(i) *F is linearly perfect.*
(ii) *$\mathbf{D}(F)$ is linearly perfect.*
(iii) *$\mathbf{U}(F)_*(\mathbf{U}(V_2,W_2))$, the underlying "complex" of $\mathbf{U}(V_1,W_1)$-modules obtained from $\mathbf{U}(V_2,W_2)$ by restriction of scalars along $\mathbf{U}(F)$, is perfect.*
(iv) *The complex $\mathbf{U}(F)^!(\mathbf{U}(V_1,W_1)) = \operatorname{RHom}_{\mathbf{U}(V_1,W_1)}(\mathbf{U}(V_2,W_2),\mathbf{U}(V_1,W_1))$ of $\mathbf{U}(V_2,W_2)$-modules is perfect.*
(v) *$\mathbf{U}(F)$ is a duality morphism of finite type (and its virtual codimension equals $-\operatorname{ind}^+(F)$).*

(vi) $\mathbf{U}(\mathbf{D}(F))$ *is a duality morphism of finite type (and its virtual codimension equals* $-\operatorname{ind}^-(F)$*).*

PROOF. Left to the reader. □

REMARK. (a) For a linearly perfect morphism F as above, the relative dualizing module of $\mathbf{U}(F)$ is

$$\omega_{\mathbf{U}(F)} = \mathrm{H}^{\mathrm{ind}^+(F)}(\mathbf{U}(F)^!(\mathbf{U}(V_1, W_1))) \quad \text{according to (7.6.3)}$$

which yields by a simple computation: **104**

$$\omega_{\mathbf{U}(F)} = \mathbf{U}(V_2, W_2) \otimes_K \det(F)\,.$$

As in (8.5.7) (v), we define the *canonical module* of $\mathbf{U}(V, W)$ to be

$$\omega_{\mathbf{U}(V,W)} = \mathbf{U}(V, W) \otimes_K \det(V, W)\,,$$

so that one has as usual

- for $W = 0$ the ordinary canonical module $\omega_{\mathrm{Sym}(V)} = \mathrm{Sym}(V) \otimes_K \det(V)$, which may and will be identified with the module Ω^v_V (where $v = \dim_K V$) of differential forms of maximal degree over $\mathrm{Sym}(V)$,
- for a linearly perfect morphism F as before,

$$\omega_{\mathbf{U}(V_2,W_2)} = \mathrm{H}^{\mathrm{ind}^+(F)}(\mathbf{U}(F)^!(\omega_{\mathbf{U}(V_1,W_1)}))\,.$$

If the graded vectorspace (V, W) is to be understood we will set

$$\mathbf{U} := \mathbf{U}(V, W) \quad \text{and} \quad \mathbf{UD} := \mathbf{U}(\mathbf{D}(V, W)) = \mathbf{U}(W^*, V^*)\,,$$

and similarly we write simply $\omega_{\mathbf{U}}$ and $\omega_{\mathbf{UD}}$ for the corresponding canonical modules.

If we need to separate the two degrees, we will write $\mathbf{U}_i^j$ for the homogeneous component $\mathrm{Sym}_i(V) \otimes_K \bigwedge^j(W)$ of bidegree (i, j). Analogously, for any graded $\mathbf{U}$-module $N = \bigoplus N_i^j$, N_i^j denotes the component of degree (i, j).

The key observation by J. Bernšteĭn, S. I. Gelfand and I. M. Gelfand in [**BGG78**], slightly generalised, was now that homological algebra over $\mathbf{U}$ or $\mathbf{UD}$ is essentially the "same"—but by a non-trivial equivalence.

To formulate it precisely, let us (re-)introduce the following module categories:

- $\mathrm{Mod}_{lf}(\mathbf{U})$ will denote the category of graded $\mathbf{U}$-modules N which are *locally finite for the total degree*, that is, for any integer k, the K-vectorspace

$$(\int N)_k := \bigoplus_{i+j=k} N_i^j$$

  is finite dimensional. In particular, each component N_i^j is of finite dimension over K.

As $\mathbf{U}$ itself is certainly locally finite for the total degree and as it is furthermore **105**
"positively graded", namely concentrated in degrees $\mathbb{N} \times [0, \dim_K W]$, it follows that

- $\mathrm{mod}(\mathbf{U})$, the category of all finitely generated graded $\mathbf{U}$-modules, is a full subcategory of $\mathrm{Mod}_{lf}(\mathbf{U})$, whose objects N satisfy:
  - For all but finitely many j, $N^j = 0$ and
  - For all but finitely many negative i, $N_i = 0$.

  In particular, the associated $\mathbb{Z}$-graded module $\int N$ is essentially positively graded.

- art($\mathbf{U}$) will denote the full subcategory of mod($\mathbf{U}$) which consists of all graded artinian $\mathbf{U}$-modules N. Equivalently, each component N_i^j is finite dimensional over K and there are only finitely many pairs (i,j) such that $N_i^j \neq 0$. The artinian modules constitute a Serre-subcategory of mod($\mathbf{U}$), so we may form the (abelian) quotient category
- Proj-$\mathbf{U}$:= mod($\mathbf{U}$)/ art($\mathbf{U}$). Forgetting the $\bigwedge(W)$-structure, objects in it can be itentified with $\mathbb{Z}$-graded, coherent sheaves of modules on the projective space $\mathbb{P}(V)$, the action of the exterior algebra being recovered through the $\mathcal{O}_{\mathbb{P}(V)}$-linear maps which represent the multiplication with the "1-forms" in W.

  The "typical object" in Proj-$\mathbf{U}$ is hence a finite family $(F^j)_j$ of coherent sheaves of $\mathcal{O}_{\mathbb{P}(V)}$-modules together with $\mathcal{O}_{\mathbb{P}(V)}$-linear maps

$$\lambda^j \colon F^{j-1} \otimes_K W \longrightarrow F^j$$

  satisfying $\lambda^{j+1}(\lambda^j(f \otimes w) \otimes w) = 0$ for all local sections f of F^{j-1} and any w in W.

  In still different terms, Proj-$\mathbf{U}$ can be identified as the category of coherent sheaves of modules over the topological space $\mathbb{P}(V)$, ringed by the $\mathbb{Z}$-graded sheaf $\mathcal{O}_{\mathbb{P}(V)} \otimes_K \bigwedge(W)$.

  We will denote the projection functor from mod($\mathbf{U}$) to Proj-$\mathbf{U}$ by $\mathbf{a}$ and think of it as "sheafification" as in the classical case where $W = 0$.

Passing now to the underlying $\mathbb{Z}/2\mathbb{Z}$-graded structures, we are in the realm of "super-geometry":

$\mathbf{U} = \mathrm{Sym}(V) \otimes_K (\bigwedge^+(W) \oplus \bigwedge^-(W))$ is the ($\mathbb{Z} \times \mathbb{Z}/2\mathbb{Z}$-graded) coordinate ring of the affine super-space $\mathbf{A}(V,W)$ attached to the linear super-space (V,W), the projective space $\mathbb{P}(V)$ ringed by the $\mathbb{Z}/2\mathbb{Z}$-graded sheaf of K-algebras $\mathcal{O}_{\mathbb{P}(V)} \otimes_K$
106 $(\bigwedge^+(W) \oplus \bigwedge^-(W))$ representing the underlying projective super-space $\mathbb{P}(V,W)$.

REMARK. (b) $\mathbb{P}(V,W)$ is actually the "algebraic" projective super-space as the structure sheaf on its underlying reduced scheme, which is the usual projective space $\mathbb{P}(V)$, is given by the (germs of) algebraic—rather than holomorphic—functions. Observe that we did not restrict our field of coefficients! For more details in the real or complex case see for example [**Lei83**].

Denoting $\mathcal{O}_{\mathbb{P}(V,W)}$ the just introduced structure sheaf on $\mathbb{P}(V,W)$ and mod($\mathbb{P}(V,W)$) the category of coherent sheaves of modules over it, by definition the sheaf cohomology of a coherent $\mathbb{Z}/2\mathbb{Z}$-graded $\mathcal{O}_{\mathbb{P}(V,W)}$-module $\underline{F}$ is the $\mathbb{Z} \times \mathbb{Z}/2\mathbb{Z}$-graded K-vectorspace which is given in degree $(i,\underline{j})$ by

$$\mathrm{H}^i(\underline{F})_{\underline{j}} = \mathrm{Ext}^i_{\mathrm{mod}(\mathbb{P}(V,W))}(\mathcal{O}_{\mathbb{P}(V,W)}, \underline{F}(\underline{j}))\,.$$

If now $(F^j)_j$ is an object in Proj-$\mathbf{U}$ as above, set $\underline{F} = F^+ \oplus F^-$, with *even* component $F^+ = \bigoplus_j F^{2j}$ and *odd* component $F^- = \bigoplus_j F^{2j+1}$.

It represents naturally an object in mod($\mathbb{P}(V,W)$). For any two objects $F = (F^j)_j$ and $G = (G^j)_j$ in Proj-$\mathbf{U}$ one has then obviously

$$\mathrm{Ext}^i_{\mathrm{mod}(\mathbb{P}(V,W))}(\underline{F}, \underline{G})_{\underline{j}} = \bigoplus_{l = \underline{j} \bmod 2} \mathrm{Ext}^i_{\mathrm{Proj}\text{-}\mathbf{U}}(F, G(l))$$

and as $\underline{\mathbf{a}(\mathbf{U})} = \mathscr{O}_{\mathbb{P}(V,W)}$, in particular

$$\mathrm{H}^i(\underline{F})_{\underline{j}} = \bigoplus_{l=\underline{j} \bmod 2} \mathrm{Ext}^i_{\mathrm{Proj}\text{-}\mathbf{U}}(\mathbf{a}(\mathbf{U}), F(l))\,.$$

Hence, passing to the underlying $\mathbb{Z}/2\mathbb{Z}$-grading yields an exact embedding of $\mathrm{Proj}\text{-}\mathbf{U}$ into $\mathrm{mod}(\mathbb{P}(V,W))$ and in this sense, the study of objects in $\mathrm{Proj}\text{-}\mathbf{U}$ equals the study of special sheaves on projective super-space, namely of those whose $\mathbb{Z}/2\mathbb{Z}$-grading can be "refined" to a $\mathbb{Z}$-grading. Similar remarks apply to the corresponding natural functor from $\mathrm{mod}(\mathbf{U})$ into $\mathrm{mod}(\mathbf{A}(V,W))$, the category of graded coherent sheaves of modules on the affine "super-cone" over $\mathbb{P}(V,W)$.

In all, we have a commutative diagram of abelian categories and exact functors between them

$$\begin{array}{ccc}
\mathrm{mod}(\mathbf{U}) & \xrightarrow{\text{underlying } \mathbb{Z}/2\mathbb{Z}\text{-grading}} & \mathrm{mod}(\mathbf{A}(V,W)) \\
\Big\downarrow & \longleftarrow \text{"sheafification" } \mathbf{a}(-) \longrightarrow & \Big\downarrow \\
\mathrm{Proj}\text{-}\mathbf{U} & \xrightarrow{\text{underlying } \mathbb{Z}/2\mathbb{Z}\text{-grading}} & \mathrm{mod}(\mathbb{P}(V,W))
\end{array}$$

so that one may be tempted to think of $\mathbf{A}(V,W)$ as the *even affine cone*, of $\mathbf{U}$ as 107
the "homogeneous coordinate ring" of the "*total cone*" and of $(\mathbb{P}(V), \mathbf{a}(\mathbf{U}))$ as the "*odd projective cone*" associated to the projective super-space defined by (V,W).

Coming back to our original subject, recall that $\underline{\mathrm{MCM}}(\mathbf{U})$ denotes the stable category of graded maximal Cohen–Macaulay modules over $\mathbf{U}$.

Deriving the abelian categories introduced, we are led to study

9.3 The associated triangulated categories. To investigate various triangulated categories canonically attached to a linear super-space (V,W), we will need the following criterion observed by A. A. Beĭlinson:

LEMMA AND DEFINITION 9.3.1. [**Beĭ84**, Lemma 1]

(i) *Let* C *be any triangulated category,* $(X_i)_i$ *a family of objects in* C*. Then there is a smallest full and triangulated subcategory of* C *which contains this family of objects. Let us call it the triangulated span of* $(X_i)_i$ *in* C*. If this category is already equivalent to* C *with respect to the embedding, we say that* $(X_i)_i$ *generates* C *(as a triangulated category).*[1]

(ii) *Let* $F\colon \mathsf{C} \to \mathsf{D}$ *be an exact functor between triangulated categories. Then the restriction of* F *to the subcategory spanned by* $(X_i)_i$ *is full, faithful or an exact equivalence onto the subcategory of* D *which is spanned by* $(F(X_i))_i$ *if the maps*

$$F^k(X_i, X_j)\colon \mathrm{Hom}_{\mathsf{C}}(X_i, X_j[k]) \longrightarrow \mathrm{Hom}_{\mathsf{D}}(F(X_i), F(X_j)[k])$$

are injective, surjective or isomorphisms[2] *(of abelian groups) for all* k *and any pair* (X_i, X_j) *of objects in the family.* □

REMARK. (a) If the triangulated category C carries further gradings, say defined by shift functors $-(\delta)$, a family $(X_i)_i$ of objects in C will be said *to generate*

[1] In current terminology, generation allows for taking direct summands as well.

[2] Beĭlinson only proves the statement for isomorphisms. This does not hold for injectivity or surjectivity, and those statements are not needed later.

(a subcategory of) C *up to shifts*, if the family $(X_i(\delta))_{i,\delta}$ generates it in the aforementioned sense.

As now $\mathbf{U}$ is already bigraded, its derived categories will be triply graded by the functors $-[k](i,j)$, with k, i, j any integers.

What now simplifies considerably the homological algebra over $\mathbf{U}$ is the fact
108 that most interesting triangulated categories attached to it can be generated up to shift by single modules—considered as complexes in the usual way. A formal indication why this makes life easier is obtained by just considering the associated Grothendieck groups:

REMARK. (b) If C is any triangulated category which can be generated by a family of objects $(X_i)_i$, the associated Grothendieck group $\mathsf{K}_0(\mathsf{C})$ is generated as an abelian group by the classes of these objects. If now C carries further gradings, defined by shift functors $-(\delta)$, δ in some group Δ of degrees, then $\mathsf{K}_0(\mathsf{C})$ becomes in a natural way a module over the integral group ring of Δ and if the family $(Y_j)_j$ generates C up to shifts, the classes $\mathrm{cl}(Y_j)$ generate $\mathsf{K}_0(\mathsf{C})$ as such a module.

Hence if C can be generated by a single object (up to shifts), its Grothendieck group is a cyclic group (resp. module over the group ring).

In case of $\mathbf{U}$, some naturally occurring triangulated categories are the following ones:

(9.3.2) $\mathrm{D}^{\mathrm{b}}_{\mathrm{art}}(\mathbf{U})$, the full triangulated subcategory of $\mathrm{D}^{\mathrm{b}}(\mathbf{U})$ whose objects are those complexes with artinian cohomology, is equivalent to $\mathrm{D}^{\mathrm{b}}(\mathrm{art}(\mathbf{U}))$.[3]

(i) It can be obtained as the thick hull of the image of the forgetful functor

$$\mathbf{U}(p_W)_*\colon \mathrm{D}^{\mathrm{b}}(\textstyle\bigwedge(W)) \longrightarrow \mathrm{D}^{\mathrm{b}}(\mathbf{U}),$$

associated to the projection $p_W\colon (V,W) \to (0,W)$ onto the odd quotient space. (Of course, it is also the thick hull of the image of the forgetful functor associated to the zero-map $0\colon (V,W) \to (0,0)$, but this map is not linearly perfect, whence we prefer p_W!).

(ii) $\mathrm{D}^{\mathrm{b}}_{\mathrm{art}}(\mathbf{U})$ is generated up to shifts by the single $\mathbf{U}$-module K.

These assertions are obvious.

(iii) For the Grothendieck groups one has hence:

$$\mathsf{K}_0(\mathrm{D}^{\mathrm{b}}_{\mathrm{art}}(\mathbf{U})) = \mathsf{K}_0(\mathrm{D}^{\mathrm{b}}(\mathrm{art}(\mathbf{U}))) = \mathsf{K}_0(\mathrm{art}(\mathbf{U})) = \mathrm{cl}(K)\cdot\mathbb{Z}[s,s^{-1},\lambda,\lambda^{-1}],$$

with $\mathrm{cl}(K(i,j)) = (-1)^j\,\mathrm{cl}(K)s^i\lambda^j$. (The sign has been introduced just to keep track of the "parity" of even and odd parts. It makes formulas look more natural.)

(iv) From the explicit knowledge of the natural projective resolution of K over $\mathbf{U}$, (8.5.7), we get:

109 $$\bigoplus_{i,j}\mathrm{RHom}_{\mathbf{U}}(K,K(i,j)) = \bigoplus_{i,j}\mathrm{Sym}_i(W^*)\otimes_K \textstyle\bigwedge^j(V^*)[-i-j](j,i)$$

as complexes of bigraded K-vectorspaces.

Accordingly, $\mathrm{Ext}^k_{\mathbf{U}}(K,K)_{(i,j)} = 0$ for $k+i+j\neq 0$ and

$$\mathrm{Ext}^{i+j}_{\mathbf{U}}(K,K)_{(-j,-i)} = \mathrm{Sym}_i(W^*)\otimes_K \textstyle\bigwedge^j(V^*) = \mathbf{UD}^j_i.$$

[3] For details see Remark A.9.1.

The multiplicative structure on the Yoneda-Ext-algebra is precisely the *opposite* one of **UD**, due to our conventions on shifts, (8.4), and our preference for right modules. As $\bigwedge(V^*)^{\mathrm{op}}$ is canonically isomorphic as a K-algebra to $\bigwedge(V^*)$ and $\mathrm{Sym}(W^*)$ is commutative, the Yoneda-Ext-algebra of K over $\mathbf{U}$ "is" $\mathbf{U}(W^*, V^*)$—but regraded.

(9.3.3) $\mathrm{D}^{\mathrm{b}}_{\mathrm{perf}}(\mathbf{U})$ denotes, as always, the full triangulated subcategory of graded perfect complexes in $\mathrm{D}^{\mathrm{b}}(\mathbf{U})$.

(i) It can be obtained as the thick hull of either the image of the left adjoint $\mathbf{U}(i_V)^*$ or the right adjoint $\mathbf{U}(i_V)^!$ of the forgetful functor associated to the inclusion $i_V\colon (V,0) \to (V,W)$ of the even subspace. These functors

$$\mathbf{U}(i_V)^*, \mathbf{U}(i_V)^!\colon \mathrm{D}^{\mathrm{b}}(\mathrm{Sym}(V)) \longrightarrow \mathrm{D}^{\mathrm{b}}(\mathbf{U}),$$

are well defined by (9.2.1) as i_V is linearly perfect and their images span the perfect subcategory as $\mathrm{Sym}(V)$ is of finite global dimension, see (1.3) (b) and remark that

$$\mathbf{U}(i_V)^!(-) = \mathbf{U}(i_V)^*(-) \otimes_{\mathbf{U}} \omega_{\mathbf{U}/\,\mathrm{Sym}(V)} = \mathbf{U}(i_V)^*(-) \otimes_K (\det(W))^{-1},$$

by Remark (a) in (9.2).

(ii) According to these choices, there are—at least—two canonical generators for $\mathrm{D}^{\mathrm{b}}_{\mathrm{perf}}(\mathbf{U})$ imposing themselves, namely either $\mathbf{U} = \mathbf{U}(i_V)^*(\mathrm{Sym}(V))$ or else $\omega_{\mathbf{U}} = \mathbf{U}(i_V)^!(\omega_{\mathrm{Sym}(V)})$.

(iii) For the Grothendieck group, remark that a finitely generated projective and graded $\mathbf{U}$-module is necessarily a direct sum of modules of the form $\mathbf{U}(i,j)$ for various (i,j). It follows that

$$\mathsf{K}_0(\mathrm{D}^{\mathrm{b}}_{\mathrm{perf}}(\mathbf{U})) = \mathsf{K}_0(\mathbf{U}) = \mathrm{cl}(\mathbf{U}) \cdot \mathbb{Z}[s, s^{-1}, \lambda, \lambda^{-1}],$$

and that, using the same conventions as in ((9.3.2)) (iii), one has[4] **110**

$$\mathrm{cl}(\omega_{\mathbf{U}}) = \mathrm{cl}(\mathbf{U} \otimes_K \det(V,W)) = \mathrm{cl}(\mathbf{U}(-v,w)) = \mathrm{cl}(\mathbf{U}) \cdot (-1)^w s^{-v} \lambda^w .$$

(iv) As concerns the graded endomorphism rings of these generators, one obtains as in (8.4):

$$\bigoplus_{i,j} \mathrm{RHom}_{\mathbf{U}}(\mathbf{U}, \mathbf{U}(i,j)) = \bigoplus_{i,j} \mathrm{RHom}_{\mathbf{U}}(\omega_{\mathbf{U}}, \omega_{\mathbf{U}}(i,j)) = \mathbf{U}^{\mathrm{op}},$$

whence the corresponding Ext-algebra is isomorphic to $\mathbf{U}^{\mathrm{op}}$, again conveniently regraded. By the same argument as in ((9.3.2)) (iv), $\mathbf{U}^{\mathrm{op}}$ is isomorphic as a graded K-algebra to $\mathbf{U}(V,W)$.

(9.3.4) $\mathrm{D}^{\mathrm{b}}_{\mathrm{a/p}}(\mathbf{U})$ will denote the full triangulated subcategory of all perfect complexes with artinian cohomology in $\mathrm{D}^{\mathrm{b}}(\mathbf{U})$. Hence it is the intersection of $\mathrm{D}^{\mathrm{b}}_{\mathrm{art}}(\mathbf{U})$ and $\mathrm{D}^{\mathrm{b}}_{\mathrm{perf}}(\mathbf{U})$.

(i) Accordingly, it can be obtained as the thick hull of the image of $\mathrm{D}^{\mathrm{b}}_{\mathrm{perf}}(\bigwedge(W))$ under $\mathbf{U}(p_W)_*$ or also as the thick hull of the image of $\mathrm{D}^{\mathrm{b}}_{\mathrm{art}}(\mathrm{Sym}(V))$ under either functor $\mathbf{U}(i_V)^*$ or $\mathbf{U}(i_V)^!$.

---

[4]Here and in the sequel $v = \dim_K V$ and $w = \dim_K W$.

(ii) It follows that this category can be generated up to shifts by either the single module

$$\bigwedge(W) = \mathbf{U}(p_W)_*(\bigwedge(W)) = \mathbf{U}(i_V)^*(K)$$

or also

$$\omega_{\bigwedge(W)} = \bigwedge(W) \otimes_K (\det(W))^{-1} = \mathbf{U}(i_V)^!(K)\,,$$

as K (resp. $\bigwedge(W)$) generate $\mathrm{D}^{\mathrm{b}}_{\mathrm{art}}(\mathrm{Sym}(V))$, (respectively $\mathrm{D}^{\mathrm{b}}_{\mathrm{perf}}(\bigwedge(W))$).

(iii) For the Grothendieck group one finds hence

$$\mathsf{K}_0(\mathrm{D}^{\mathrm{b}}_{\mathrm{a/p}}(\mathbf{U})) = \mathrm{cl}(\bigwedge(W)) \cdot \mathbb{Z}[s, s^{-1}, \lambda, \lambda^{-1}]\,,$$

and, filtering $\bigwedge(W)$ by the powers of its natural augmentation ideal, it follows that the image of the generator $\mathrm{cl}(\bigwedge(W))$ in $\mathsf{K}_0(\mathrm{D}^{\mathrm{b}}_{\mathrm{art}}(\mathbf{U}))$ is given by

$$\begin{aligned}\mathrm{cl}(\bigwedge(W)) &= \mathrm{cl}(\bigoplus_j \bigwedge^j(W)) = \mathrm{cl}(\bigoplus_j K(0,-j)^{\oplus \binom{w}{j}})\\ &= \mathrm{cl}(K)\cdot(1-\lambda^{-1})^w\,.\end{aligned}$$

As $\bigwedge(W)$ over $\mathbf{U}$ can be resolved by an appropriate Koszul complex, one obtains for the class of $\bigwedge(W)$ in $\mathsf{K}_0(\mathrm{D}^{\mathrm{b}}_{\mathrm{perf}}(\mathbf{U})) = \mathsf{K}_0(\mathbf{U})$:

$$\begin{aligned}\mathrm{cl}(\bigwedge(W)) &= \sum_i (-1)^i\,\mathrm{cl}(\mathbf{U}\otimes_K \bigwedge^i(V))\\ &= \sum_i (-1)^i\,\mathrm{cl}(\mathbf{U}(i,0)^{\oplus\binom{v}{i}}) = \mathrm{cl}(\mathbf{U})\cdot(1-s^{-1})^v\,.\end{aligned}$$

111 As $\mathrm{cl}(\omega_{\bigwedge(W)}) = \mathrm{cl}(\bigwedge(W)(0,w)) = \mathrm{cl}(\bigwedge(W))\cdot(-\lambda)^w$, the above formulas, if rewritten exclusively using dualizing modules, become:

$$\mathrm{cl}(\omega_{\bigwedge(W)}) = \mathrm{cl}(K)\cdot(1-\lambda)^w = \mathrm{cl}(\omega_{\mathbf{U}})\cdot(s-1)^v\,.$$

(iv) Using once again the resolution of $\bigwedge(W)$ over $\mathbf{U}$ by a Koszul complex, one obtains

$$\begin{aligned}\bigoplus_{i,j}\mathrm{RHom}_{\mathbf{U}}(\bigwedge(W),\bigwedge(W)(i,j)) &= \bigoplus_{i,j}\mathrm{RHom}_{\mathrm{Sym}(V)}(K,\bigwedge(W)(i,j))\\ &= \bigoplus_i \bigwedge(W)\otimes_K \bigwedge^i(V)^*[-i](i,0)\,,\end{aligned}$$

so that the Yoneda-Ext-algebra of $\bigwedge(W)$ over $\mathbf{U}$ is easily seen to be isomorphic to $\bigwedge(W)\otimes_K(\bigwedge(V)^*)^{\mathrm{op}} = \bigwedge(W)\otimes_K\bigwedge(V^*)$[5] using [**Bou70**, III, §11.5]—with the same change in grading as in the other cases considered.

Of course, the result for $\omega_{\bigwedge(W)}$ is the same.

[5]This is clear as vectorspaces, but the algebra structure requires justification. One uses the fact that the Koszul resolution of $\bigwedge(W)$ has a natural action of $\bigwedge(W)\otimes_K\bigwedge(V^*)$.

(9.3.5) Now let us consider $\mathrm{D^b}(\mathbf{U})$ itself.

(i) It can be generated up to shifts by either single module $\mathrm{Sym}(V)$ or $\omega_{\mathrm{Sym}(V)}$. For this, remark that any finitely generated graded $\mathbf{U}$-module admits a finite filtration by submodules such that the associated subquotients are annihilated by $W = \mathrm{Sym}_0(V) \otimes_K \bigwedge^1(W)$ in $\mathbf{U}$.

Hence these subquotients are naturally $\mathrm{Sym}(V)$-modules and, as the polynomial ring $\mathrm{Sym}(V)$ is of finite global dimension, they admit finite resolutions by graded free $\mathrm{Sym}(V)$-modules. This shows that $\mathrm{Sym}(V)$ as a $\mathbf{U}$-module generates $\mathrm{mod}(\mathbf{U})$, and hence also $\mathrm{D^b}(\mathrm{mod}(\mathbf{U}))$.

As $\omega_{\mathrm{Sym}(V)}$ is a faithfully projective $\mathrm{Sym}(V)$-module, it generates as well.

(ii) The Grothendieck group of $\mathrm{D^b}(\mathbf{U})$ equals $\mathsf{G}_0(\mathbf{U})$—following [**Qui73**]. It is freely generated over $\mathbb{Z}[s, s^{-1}, \lambda, \lambda^{-1}]$ by the class of $\mathrm{Sym}(V)$.

To see this, recall first that by Remark (b) above, this class necessarily generates the Grothendieck group. That there is no non-trivial relation follows from the obvious fact that $\mathbf{U}(i_V)_*$, the forgetful functor associated to i_V, "preserves" the class of $\mathrm{Sym}(V)$, which is known to generate freely $\mathsf{G}_0(\mathrm{Sym}(V))$ by [**Qui73**, Thm. 6].

(iii) Using the natural projective resolution of K over $\bigwedge(W)$, exhibited in (8.5.7), one obtains

$$\bigoplus_{i,j} \mathrm{RHom}_{\mathbf{U}}(\mathrm{Sym}(V), \mathrm{Sym}(V)(i,j)) = \bigoplus_{i,j} \mathrm{RHom}_{\bigwedge(W)}(K, \mathrm{Sym}(V)(i,j))$$
$$= \bigoplus_{i} \mathrm{Sym}(V) \otimes_K \Gamma^i(W)^*[-i](0,i)$$ **112**
$$= \bigoplus_{i} \mathrm{Sym}(V) \otimes_K \mathrm{Sym}_i(W^*)[-i](0,i)\,.$$

Hence the Ext-algebra of $\mathrm{Sym}(V)$—or also $\omega_{\mathrm{Sym}(V)}$—over $\mathbf{U}$ is given by

$$\mathrm{Sym}(V) \otimes_K \mathrm{Sym}(W^*) = \mathrm{Sym}(V \oplus W^*)$$

regraded as always.

To state that our main theorem below, it is useful to consider still another generator of $\mathrm{D^b}(\mathbf{U})$, namely the "true" dualizing complex of $\mathrm{Sym}(V)$, which is given by $\mathbf{K}_V = \omega_{\mathrm{Sym}(V)}[v]$. Its graded endomorphism ring in the derived category $\mathrm{D^b}(\mathbf{U})$ is still the same as that of $\mathrm{Sym}(V)$, and its class in the Grothendieck group $\mathsf{G}_0(\mathbf{U}) = \mathsf{K}_0(\mathrm{D^b}(\mathbf{U}))$ is apparently

$$\mathrm{cl}(\mathbf{K}_V) = \mathrm{cl}(\mathrm{Sym}(V)[v](-v,0)) = \mathrm{cl}(\mathrm{Sym}(V)) \cdot (-s)^{-v}\,.$$

Remark that $\mathbf{K}_V = 0_V^!(K)$, $0_V\colon (0,0) \to (V,0)$ the zero map considered also as the structure morphism from the affine space $\bigwedge(V)$ to the point $\mathrm{Spec}(K)$, $0_V^!$ is the usual functor from the duality theory for smooth morphisms, [**Har66**, III].

To finish consideration of $\mathrm{D^b}(\mathbf{U})$ for the moment, observe still how the classes of $\mathbf{U}$, $\bigwedge(W)$ and K are expressed in $\mathsf{G}_0(\mathbf{U})$—using the same arguments as in ((9.3.4)) (iii):

$$\mathrm{cl}(\mathbf{U}) = \mathrm{cl}(\mathrm{Sym}(V)) \cdot (1-\lambda^{-1})^w\,,$$
$$\mathrm{cl}(K) = \mathrm{cl}(\mathrm{Sym}(V)) \cdot (1-s^{-1})^v\,, \quad \text{and}$$
$$\mathrm{cl}(\textstyle\bigwedge(W)) = \mathrm{cl}(\mathrm{Sym}(V)) \cdot (1-\lambda^{-1})^w(1-s^{-1})^v\,.$$

Now let us name and identify the quotient categories of $\mathrm{D^b}(\mathbf{U})$ which are obtained by factoring out the various thick subcategories introduced.

(9.3.6) $\mathrm{D^b}(\mathbf{U})/\,\mathrm{D^b_{art}}(\mathbf{U})$ is naturally equivalent to $\mathrm{D^b}(\mathrm{Proj}\text{-}\mathbf{U})$. Hence this derived category of Proj-**U** can be generated—up to shifts—by the single object $\mathbf{a}(\mathrm{Sym}(V))$, the sheafification of the **U**-module $\mathrm{Sym}(V)$. It follows that the Grothendieck group of Proj-(**U**) is given by

$$\mathsf{K}_0(\mathrm{Proj}\text{-}\mathbf{U}) = \mathsf{K}_0(\mathrm{D^b}(\mathrm{Proj}\text{-}\mathbf{U})) = \mathsf{K}_0(\mathrm{D^b}(\mathbf{U}))/\mathsf{K}_0(\mathrm{D^b_{art}}(\mathbf{U}))\,,$$

113 whence $\mathsf{K}_0(\mathrm{Proj}\text{-}\mathbf{U}) = \mathrm{cl}(\mathbf{a}(\mathrm{Sym}(V)))\cdot\mathbb{Z}[s,s^{-1},\lambda,\lambda^{-1}]/(1-s^{-1})^v$.

REMARK. (c) Passing to the projective super-space $\mathbb{P}(V,W)$, one may deduce that its Grothendieck group is obtained from the above by adding the relation $\lambda^2 = 1$. Remark also that the sheaf on $\mathbb{P}(V,W)$ which is the image of $\mathbf{a}(\mathrm{Sym}(V))$ is just the structure sheaf of the underlying reduced space $\mathbb{P}(V,W)_{\mathrm{red}}$, which in turn is nothing but ordinary projective space $\mathbb{P}(V)$. Hence, the structure sheaf of the underlying reduced space generates the Grothendieck group, but not the structure sheaf itself, which is the image of $\mathbf{a}(\mathbf{U})$. More precisely, one has in that group $\mathsf{K}_0(\mathbb{P}(V,W))$:

$$\begin{aligned}\mathrm{cl}(\mathcal{O}_{\mathbb{P}(V,W)}) &= \mathrm{cl}(\mathcal{O}_{\mathbb{P}(V)})\cdot(1-\lambda^{-1})^w\\ &= \mathrm{cl}(\mathcal{O}_{\mathbb{P}(V)})\cdot 2^w(1-\lambda) \bmod (\lambda^2-1)\,.\end{aligned}$$

This shows a significant difference between "ordinary" and "super" projective geometry. This explains, for example, the phenomenon observed by O. V. Ogievetskiĭ and I. B. Penkov, [**OP84**], that, extending naturally the theory of Serre duality of projective super-manifolds, the top cohomology group $\mathrm{H}^v(\omega_{\mathbb{P}(V,W)})$ of the dualizing sheaf $\omega_{\mathbb{P}(V,W)}$—which is just the sheaf obtained from $\omega_{\mathbf{U}}$—carries a natural trace, but is not one-dimensional. In fact, the "natural" domain of definition for the trace is the group

$$\mathrm{Ext}^v_{\mathbb{P}(V,W)}(\mathcal{O}_{\mathbb{P}(V)},\omega_{\mathbb{P}(V)}) = \mathrm{H}^v(\mathbb{P}(V),\omega_{\mathbb{P}(V)}) = K\,.$$

Rephrasing the above, the image of **U** in Proj-**U**—or $\mathrm{mod}(\mathbb{P}(V,W))$—generates only the "perfect" coherent sheaves:

(9.3.7) $\mathrm{D^b_{perf}}(\mathrm{Proj}\text{-}\mathbf{U}) := \mathrm{D^b_{perf}}(\mathbf{U})/\,\mathrm{D^b_{a/p}}(\mathbf{U})$ can be identified—up to an exact equivalence—with the full triangulated subcategory of $\mathrm{D^b}(\mathrm{Proj}\text{-}\mathbf{U})$ which is generated up to shifts by $\mathbf{a}(\mathbf{U})$, the structure sheaf of the ringed space associated to Proj-**U**. It is also the thick hull of the image of either induced functor

$$i_V^*, i_V^!\colon \mathrm{D^b}(\mathbb{P}(V)) \longrightarrow \mathrm{D^b}(\mathrm{Proj}\text{-}\mathbf{U})\,.$$

It represents hence those objects in $\mathrm{D^b}(\mathrm{Proj}\text{-}\mathbf{U})$ which are obtainable from the
114 underlying (ordinary) projective space $\mathbb{P}(V)$ by "lifting". Accordingly, we set

(9.3.8) <u>Prim</u>(**U**) $= \mathrm{D^b}(\mathrm{Proj}\text{-}\mathbf{U})/\,\mathrm{D^b_{perf}}(\mathrm{Proj}\text{-}\mathbf{U})$, and call it the category of (classes of) *primitive objects* over Proj-**U**, in analogy to the theory of primitive cycles for projectives schemes.

This quotient category is still generated up to shifts by the class of $\mathbf{a}(\mathrm{Sym}(V))$. To abbreviate notations, let σ denote the class of this object in the Grothendieck group, to get

$$\begin{aligned}\mathsf{K}_0(\underline{\mathrm{Prim}}(\mathbf{U})) &= \sigma \cdot \mathbb{Z}[s, s^{-1}, \lambda, \lambda^{-1}]/(1-s^{-1})^v, (1-\lambda^{-1})^w \\ &= \sigma \cdot \mathbb{Z}[h, g]/h^v, g^w\end{aligned}$$

where we set $h = 1 - s^{-1}$, which is the class of a hyperplane in the underlying (reduced or "even") space $\mathbb{P}(V)$ of "geometric K-rational points", (9.1) (i), whereas $g = 1 - \lambda^{-1}$ represents the class of a "line" in the odd quotient space W—a corresponding module of class g is given by $\mathrm{Sym}(V) \otimes_K \bigwedge(W)/(W')$, W' a hyperplane in W, which can be identified with the class of a hyperplane in $\mathbb{P}(W^*)$, the projective space of "odd K-rational points", (9.1) (ii).

Remark that this Grothendieck group is a finitely generated free abelian group of rank vw.

REMARK. (d) Passing once again to the projective super-space $\mathbb{P}(V, W)$, we can as well form the category of primitive objects whose Grothendieck group, obtained from the above by adding the relation $g(g-2) = 0$, is the direct sum of a free group of rank v and a "large" 2-torsion group.

Observe that $\underline{\mathrm{Prim}}(\mathbf{U})$ can be obtained directly from $\mathrm{D}^{\mathrm{b}}(\mathbf{U})$—up to a natural equivalence—by factoring out the thick hull of both objects $\mathbf{U}$ and K. Hence, we may also think of $\underline{\mathrm{Prim}}(\mathbf{U})$ as representing the primitive classes (of complexes) of modules over $\mathbf{U}$. Going the other way round, we could factor out first the perfect complexes, i.e. the hull of $\mathbf{U}$, to obtain—up to an exact equivalence—by Theorem (4.4.1):

(9.3.9) $\underline{\mathrm{MCM}}(\mathbf{U})$, the category of graded maximal Cohen–Macaulay modules over $\mathbf{U}$, or, equivalently, $\underline{\mathrm{APC}}(\mathbf{U})$, the homotopy category of graded complete resolutions over $\mathbf{U}$.

Remark that $\mathrm{Sym}(V)$, the generator of $\mathrm{D}^{\mathrm{b}}(\mathbf{U})$ is already an MCM over $\mathbf{U}$, (8.3.1) or (8.5.7), so that it generates $\underline{\mathrm{MCM}}(\mathbf{U})$ as a triangulated category up to shifts—there is no need to pass to the MCM-approximation first.

The natural complete resolution of $\mathrm{Sym}(V)$ over $\mathbf{U}$, described in (8.5.7), gen- **115**
erates then—up to shifts—the triangulated category $\underline{\mathrm{APC}}(\mathbf{U})$.

It follows that the Grothendieck group of MCM's over $\mathbf{U}$ modulo projectives is

$$\begin{aligned}\mathsf{K}_0(\underline{\mathrm{MCM}}(\mathbf{U})) = \mathsf{K}_0(\underline{\mathrm{APC}}(\mathbf{U})) &= \sigma \cdot \mathbb{Z}[s, s^{-1}, \lambda, \lambda^{-1}]/(1-\lambda^{-1})^w \\ &= \sigma \cdot \mathbb{Z}[s, s^{-1}, g]/g^w\end{aligned}$$

where σ and g are defined as before. (Remark that the module representing g is also already MCM, so that g keeps its original meaning.)

(9.3.10) $\underline{\mathrm{MCM}}_{\mathrm{art}}(\mathbf{U})$ will denote the thick hull in $\underline{\mathrm{MCM}}(\mathbf{U})$ of the MCM-approximation of K. It is equivalent to both the full subcategory of MCM-approximations of artinian $\mathbf{U}$-modules as well as the quotient category $\mathrm{D}^{\mathrm{b}}_{\mathrm{art}}(\mathbf{U})/\,\mathrm{D}^{\mathrm{b}}_{\mathrm{a/p}}(\mathbf{U})$.

As forming subsequent quotients of triangulated categories is—up to exact equivalences—independent of the order, it follows

PROPOSITION 9.3.11. *The quotient categories of primitive classes of complexes of coherent sheaves on* Proj -$\mathbf{U}$*,* $\underline{\mathrm{Prim}}(\mathbf{U})$*, and of MCM's modulo MCM-approximations of artinian modules,* $\underline{\mathrm{MCM}}(\mathbf{U})/\,\underline{\mathrm{MCM}}_{\mathrm{art}}(\mathbf{U})$*, are equivalent as triangulated categories.* □

To summerise the situation, we have the following commutative diagram of triangulated categories and exact functors, whose rows and columns in plain arrows are exact sequences of triangulated categories.

A dashed arrow A $\dashrightarrow$ B indicates that B is the thick hull of A under a suitable exact functor:
(9.3.12)

116

$$
\begin{array}{ccccccccc}
 & & \mathrm{D}^{\mathrm{b}}_{\mathrm{perf}}(\bigwedge(W)) & \longrightarrow & \mathrm{D}^{\mathrm{b}}(\bigwedge(W)) & \longrightarrow & \underline{\mathrm{MCM}}(\bigwedge(W)) & \longrightarrow & 0 \\
 & & \downarrow & & \downarrow & & \downarrow & & \\
\mathrm{D}^{\mathrm{b}}_{\mathrm{art}}(\mathrm{Sym}(V)) & \dashrightarrow & \mathrm{D}^{\mathrm{b}}_{\mathrm{a/p}}(\mathbf{U}) & \longrightarrow & \mathrm{D}^{\mathrm{b}}_{\mathrm{art}}(\mathbf{U}) & \longrightarrow & \underline{\mathrm{MCM}}_{\mathrm{art}}(\mathbf{U}) & \longrightarrow & 0 \\
\downarrow & & \downarrow & & \downarrow & & \downarrow & & \\
\mathrm{D}^{\mathrm{b}}(\mathrm{Sym}(V)) & \dashrightarrow & \mathrm{D}^{\mathrm{b}}_{\mathrm{perf}}(\mathbf{U}) & \longrightarrow & \mathrm{D}^{\mathrm{b}}(\mathbf{U}) & \longrightarrow & \underline{\mathrm{MCM}}(\mathbf{U}) & \longrightarrow & 0 \\
\downarrow & & \downarrow & & \downarrow & & \downarrow & & \\
\mathrm{D}^{\mathrm{b}}(\mathbb{P}(V)) & \dashrightarrow & \mathrm{D}^{\mathrm{b}}_{\mathrm{perf}}(\mathrm{Proj}\text{-}\mathbf{U}) & \longrightarrow & \mathrm{D}^{\mathrm{b}}(\mathrm{Proj}\text{-}\mathbf{U}) & \longrightarrow & \underline{\mathrm{Prim}}(\mathbf{U}) & \longrightarrow & 0 \\
\downarrow & & \downarrow & & \downarrow & & \downarrow & & \\
0 & & 0 & & 0 & & 0 & &
\end{array}
$$

Now we come to the main result of this chapter:

9.4 The Bernšteĭn–Gelfand–Gelfand correspondence. To abbreviate the formulation of the theorem, recall that the base field K can be interpreted as $\mathbf{U}(0)$, the enveloping algebra of the zero-space $0 = (0,0)$.

Passing to the cohomology of an object in $\mathrm{D}^{\mathrm{b}}(\mathbf{U}(0))$ identifies this latter category with the category of triply $\mathbb{Z}$-graded K-vectorspaces of finite total dimension.

This category is obviously generated up to shifts by the single object K, and its Grothendieck group is $\mathbb{Z}[s, s^{-1}, \lambda, \lambda^{-1}]$, the class of the graded vectorspace $K[k](i,j)$ being represented by $(-1)^{k+j} s^i \lambda^j$ in view of our general convention in ((9.3.3)) (iii).

Denote by γ the functor which regrades a triply graded vectorspace according to[6]

$$\gamma(K[k](i,j)) = K[k+i+j](-j,-i)\,.$$

This functor is exact—as $\gamma(-[k]) = \gamma(-[k](0,0)) = -[k](0,0) = -[k]$—with respect to the triangulated structure, and involutive. The induced automorphism of the Grothendieck group sends s into λ^{-1} and λ into s^{-1}.

For any given linear super-space (V,W), we will denote—in analogy to ((9.3.5)) (iii)—by

117

$$^{!}0_{(V,W)} \colon \mathrm{D}^{\mathrm{b}}(\mathbf{U}(0)) \longrightarrow \mathrm{D}^{\mathrm{b}}(\mathbf{U}(V,W))$$

the unique exact functor which commutes with the shifts $-(i,j)$ and satisfies

$$^{!}0_{(V,W)}(K) = \mathbf{K}_V = \omega_{\mathrm{Sym}(V)}[v]\,,$$

where $\mathbf{K}_V$ is considered a complex of $\mathbf{U}(V,W)$-modules as usual.

Beĭlinson's criterion (9.3.1), applied to $\mathrm{D}^{\mathrm{b}}(\mathbf{U}(V,W))$ can then be expressed as follows:

---

[6]Thus if E is a triply graded vectorspace, this means $\gamma(E)^k_{(i,j)} = E^{k+i+j}_{(-j,-i)}$.

Assume that F is any exact functor from $\mathrm{D^b}(\mathbf{U}(V,W))$ into another triangulated category. Then it is essentially uniquely determined by its composition with ${}^!0_{(V,W)}$.

If now $\mathbf{D}(V,W) = (W^*, V^*)$ is the "dual" linear super-space to (V,W), one has from ((9.3.5)) (iii):

$$\mathrm{Ext}^k_{\mathbf{U}(V,W)}(\mathbf{K}_V, \mathbf{K}_V)_{(i,j)} = \mathrm{Ext}^{k+i+j}_{\mathbf{U}(\mathbf{D}(V,W))}(\mathbf{K}_{W^*}, \mathbf{K}_{W^*})_{(-j,-i)}$$

for all integers k, i, j. The indicated vectorspaces are zero, except for $k+j=0$, in which case they equal $\mathrm{Sym}_i(V) \otimes_K \mathrm{Sym}_k(W^*)$.

In other words, the identity on $\mathrm{Sym}(V \oplus W^*)$ furnishes a natural identification of triply graded rings

$$\mathrm{Ext}_{\mathbf{U}(V,W)}(\mathbf{K}_V, \mathbf{K}_V) = \gamma(\mathrm{Ext}_{\mathbf{U}(\mathbf{D}(V,W))}(\mathbf{K}_{W^*}, \mathbf{K}_{W^*}))\,. \tag{9.4.1}$$

This is the key to:

THEOREM 9.4.2. *Let (V,W) be a linear super-space over K. Then there exists a (covariant) exact functor*

$$\mathbf{b}_{(V,W)} \colon \mathrm{D^b}(\mathbf{U}(V,W)) \longrightarrow \mathrm{D^b}(\mathbf{U}(\mathbf{D}(V,W)))$$

together with a natural isomorphism of functors on $\mathrm{D^b}(\mathbf{U}(0))$

$${}^!0_{\mathbf{D}(V,W)} \circ \gamma \xrightarrow{\simeq} \mathbf{b}_{(V,W)} \circ {}^!0_{(V,W)}\,,$$

which induces the isomorphism (9.4.1) *of triply graded rings if evaluated on the shifts and translates of K.*

The functor $\mathbf{b}_{(V,W)}$ is essentially unique, necessarily an exact equivalence and admits $\mathbf{b}_{\mathbf{D}(V,W)}$ as its inverse.

PROOF. The uniqueness of $\mathbf{b}_{(V,W)}$ follows immediately from Beĭlinson's crite- 118
rion:

$$\begin{aligned} {}^!0_{\mathbf{D}(V,W)} \circ \gamma(K[k](i,j)) &= {}^!0_{\mathbf{D}(V,W)}(K[k+i+j](-j,-i)) && \text{by definition of } \gamma, \\ &= \mathbf{K}_{W^*}[k+i+j](-j,-i) && \text{by definition of } {}^!0_{\mathbf{D}(V,W)}, \end{aligned}$$

whereas

$$\mathbf{b}_{(V,W)} \circ {}^!0_{(V,W)}(K[k](i,j)) = \mathbf{b}_{(V,W)}(\mathbf{K}_V[k](i,j))\,,$$

by definition of ${}^!0_{(V,W)}$. Hence $\mathbf{b}_{(V,W)}$ is determined on the generating family $\mathbf{K}_V(i,j)$ by the theorem, whence the uniqueness.

As its image is precisely a generating family of $\mathrm{D^b}(\mathbf{U}(\mathbf{D}(V,W)))$ and as by hypothesis the induced maps on the morphism groups are isomorphisms, it follows also that $\mathbf{b}_{(V,W)}$ is an exact equivalence. Replacing (V,W) by its "dual" $\mathbf{D}(V,W) = (W^*, V^*)$, the same criterion shows that $\mathbf{b}_{\mathbf{D}(V,W)}$ is an inverse of $\mathbf{b}_{(V,W)}$.

The non-trivial part is the *existence*. Modifying slightly the sketch of the proof in [**BGG78**], a functor $\mathbf{b}$ as claimed can be obtained in the following way, (where we suppress once again (V,W) from the notations):

First, let $N := \bigoplus N_i^j$ be any $\mathbf{U}$-module which is locally finite for the total degree. Associate to it a double complex over $\mathbf{UD}$: Its term in (complex-)bidegree (j,i) is given by $N_i^j \otimes_K \mathbf{UD}(j,i)$, and its differentials

$${}'d_i^j \colon N_i^j \otimes_K \mathbf{UD}(j,i) \longrightarrow N_i^{j+1} \otimes_K \mathbf{UD}(j+1,i)$$

and

$${}''d_i^j \colon N_i^j \otimes_K \mathbf{UD}(j,i) \longrightarrow N_{i+1}^j \otimes_k \mathbf{UD}(j,i+1)$$

are given by those **UD**-linear maps in

$$\mathrm{Hom}_{\mathbf{UD}}(N_i^j \otimes_K \mathbf{UD}(j,i), N_i^{j+1} \otimes_K \mathbf{UD}(j+1,i)) = \mathrm{Hom}_K(N_i^j, N_i^{j+1} \otimes_K W^*)$$

and

$$\mathrm{Hom}_{\mathbf{UD}}(N_i^j \otimes_K \mathbf{UD}(j,i), N_{i+1}^j \otimes_K \mathbf{UD}(j,i+1)) = \mathrm{Hom}_K(N_i^j, N_{i+1}^j \otimes_K V^*)$$

respectively, which correspond by adjunction to the action of the "odd 1-forms" in W and the action of the "even 1-forms" in V respectively on the **U**-module N:

119 The verification that $(N_i^j \otimes_K \mathbf{UD}(j,i), 'd, ''d)$ constitutes indeed a double complex is immediate. Also, a **U**-linear morphism of modules yields obviously a morphism of the corresponding double complexes.

As N is by assumption locally finite for the total degree, (which was not needed yet), the associated total complex, denoted $\mathbf{d}(N)$, consists of finitely generated, graded free **UD**-modules, such that the term in complex-degree k can be generated by homogeneous elements of degree $(-j,-i)$ with $j+i=k$. In other words, this complex is *linear* for the total degree. (Conversely, one sees immediately, that such a linear complex determines a unique **U**-module, locally finite for the total degree, of which it is the image under **d**. As furthermore two homomorphisms of linear complexes can only be homotopic if they are equal, it follows that **d** establishes in fact an equivalence of categories between $\mathrm{Mod}_{lf}(\mathbf{U})$, (9.2), and the homotopy category of linear complexes over **UD**.)

Now **d** can be extended to a functor on bounded complexes over **U** by applying it to every term and the differentials—resulting in a double complex whose "rows" are linear complexes—and then passing to the associated total complex again.

This extension is compatible with translation and carries homotopies into homotopies. As **d** transforms evidently short exact sequences of **U**-modules into short exact sequences of complexes over **UD**, it maps necessarily finite acyclic complexes over **U** into acyclic complexes over **UD** and passes hence trivially to the corresponding derived categories, so that we end up with an exact functor—again denoted **d**—from $\mathrm{D^b}(\mathrm{Mod}_{lf}(\mathbf{U}))$ into $\mathrm{D}(\mathrm{Mod}(\mathbf{UD}))$.

Now let us consider $\mathbf{d}(\mathrm{Sym}(V))$. By construction, this complex is isomorphic (as a complex of graded K-vectorspaces) to

$$(\bigoplus_i \mathrm{Sym}_i(V) \otimes_K \bigwedge(V^*)(i)[-i] \otimes_K \mathrm{Sym}(W^*), \delta)$$

which is nothing but the usual Koszul complex $(\mathrm{Sym}(V) \otimes_K \bigwedge(V^*), d)$—up to a sign in the differential—tensored with $\mathrm{Sym}(W^*)$ over K.

This shows:

(i) The inclusion of the **UD**-submodule $1 \otimes_K \det(V^*) \otimes_K \mathrm{Sym}(W^*)$ in complex-degree zero—which is isomorphic to $\mathrm{Sym}(W^*)[0](0,-v)$ as a "complex"—into $\mathbf{d}(\mathrm{Sym}(V))$ is a quasi-isomorphism of complexes.

By inspection, it is also clear that

(ii) $\mathbf{d}(-[k](i,j)) = \mathbf{d}(-)[k+i+j](-j,-i)$ for all integers k, i, j.

(iii) The induced morphisms of vectorspaces from

$$\mathrm{Hom}_{\mathrm{D^b}(\mathrm{Mod}_{lf}(\mathbf{U}))}(\mathrm{Sym}(V), \mathrm{Sym}(V)[k](i,j)) = \mathrm{Ext}^k_{\mathbf{U}}(\mathrm{Sym}(V), \mathrm{Sym}(V))_{(i,j)}$$

120 into

$$\mathrm{Ext}^{k+i+j}_{\mathbf{UD}}(\mathbf{d}(\mathrm{Sym}(V)), \mathbf{d}(\mathrm{Sym}(V)))_{(-j,-i)} = \mathrm{Ext}^{k+i+j}_{\mathbf{UD}}(\mathrm{Sym}(W^*), \mathrm{Sym}(W^*))_{(-j,-i)}$$

can be identified by ((9.3.5)) (iii), (9.4.1), as the natural isomorphism

$$\mathrm{Sym}_i(V) \otimes_K \mathrm{Sym}_k(W^*) = \mathrm{Sym}_k(W^*) \otimes_K \mathrm{Sym}_i(V) \quad \text{for } k+j=0\,,$$

and as the isomorphism of zero-objects if $k+j \neq 0$.

As $\mathbf{K}_V = \omega_{\mathrm{Sym}(V)}[v] = \mathrm{Sym}(V)[v](-v,0)$ in $\mathrm{D^b}(\mathbf{U})$, "renormalizing" $\mathbf{d}$ to

$$\mathbf{b}(-) = \mathbf{d}(- \otimes_K (\det W)^{-1}) = \mathbf{d}(-(0,w)) = \mathbf{d}(-)[w](-w,0)$$

will then yield an exact functor from $\mathrm{D^b}(\mathrm{Mod}_{lf}(\mathbf{U}))$ into $\mathrm{D}(\mathrm{Mod}(\mathbf{UD}))$ which satisfies

$$\begin{aligned}\mathbf{b}(\mathbf{K}_V) &= \mathbf{b}(\mathrm{Sym}(V)[v](-v,0))\\ &= \mathbf{d}(\mathrm{Sym}(V)[v](-v,w))\\ &= \mathbf{d}(\mathrm{Sym}(V))[w](-w,v) \quad \text{by (ii).}\end{aligned}$$

Into this complex

$$\mathbf{K}_{W^*} = \omega_{\mathrm{Sym}(W^*)}[w] = \mathrm{Sym}(W^*)[w](-w,0)$$

embeds quasi-isomorphically by (i), so that there is in fact a natural isomorphism in $\mathrm{D}(\mathrm{Mod}(\mathbf{UD}))$ from $\mathbf{K}_{W^*}$ into $\mathbf{b}(\mathbf{K}_V)$.

From (ii) and (iii) it follows that the induced morphism on the graded endomorphism ring of $\mathbf{K}_V$ coincides with the identification of (9.4.1), so that the restriction of $\mathbf{b}$ to $\mathrm{D^b}(\mathbf{U})$ satisfies the requirements of the Theorem. □

REMARK. (a) The functors $\mathbf{b}$ and $\mathbf{d}$ *do not* embed $\mathrm{D^b}(\mathrm{Mod}_{lf}(\mathbf{U}))$ into $\mathrm{D}(\mathrm{Mod}(\mathbf{UD}))$: As can be easily deduced either from the following, or the explicit construction, both functors "vanish" precisely on those complexes which are $\mathrm{Ext}_{\mathbf{U}}(K,-)$-acyclic.

Hence, the point of the theorem is essentially Nakayama's Lemma: It is equivalent—as $\mathbf{U}$ is strongly Gorenstein—to the statement that no complex X in $\mathrm{D^b}(\mathbf{U})$—which is not acyclic—can satisfy

$$\mathrm{Ext}^k_{\mathbf{U}}(K,X)_{(i,j)} = 0 \quad \text{for all integer } k,i,j\,.$$

Namely, **121**

$$\begin{aligned}\mathrm{Ext}^k_{\mathbf{U}}(K,X)_{(i,j)} &= \mathrm{Ext}^k_{\mathbf{U}}(\mathrm{RHom}_{\mathbf{U}}(X,\mathbf{U}),\mathrm{RHom}_{\mathbf{U}}(K,\mathbf{U}))_{(i,j)}\,,\\ &\qquad \text{as } \mathbf{U} \text{ is strongly Gorenstein, cf. (4.6),}\\ &\qquad \text{and graded commutative,}\\ &= \mathrm{Ext}^k_{\mathbf{U}}(\mathrm{RHom}_{\mathbf{U}}(X,\mathbf{U}),\det(V,W)^{-1}[-v])_{(i,j)}\\ &\qquad \text{by (8.5.1) (i)}\\ &= \mathrm{Ext}^{k-v}_{\mathbf{U}}(\mathrm{RHom}_{\mathbf{U}}(X,\mathbf{U}),K)_{(i+v,j-w)}\\ &\qquad \text{as } \det(V,W) = K(-v,w) \text{ by definition,}\end{aligned}$$

and this last vectorspace is canonically isomorphic to the K-dual of

$$\mathrm{Tor}^{\mathbf{U}}_{k-v}(K,\mathrm{RHom}_{\mathbf{U}}(X,\mathbf{U}))_{(-i-v,w-j)}\,.$$

As with X also $\mathrm{RHom}_{\mathbf{U}}(X,\mathbf{U})$ is a non-acyclic object in $\mathrm{D^b}(\mathbf{U})$, the Lemma of Nakayama applies.

To identify more precisely the image of $\mathrm{mod}(\mathbf{U})$ under $\mathbf{b}$ or $\mathbf{d}$, we make the

DEFINITION 9.4.3. Let C be any complex of finitely generated, graded free modules over $\mathbf{U}$ (or $\mathbf{UD}$). Then

(i) C is *almost linear* (for the total degree), if there exists an integer l such that C^{-k} can be generated by homogeneous elements of bidegree (i, j) with $k - l \leq i + j \leq k + l$ for all k.
 In this case, we say that the *deviation* (from linearity) of C is (uniformly) bounded by l.
(ii) C is a (generalised) *monad* if it is an almost linear complex which is bounded below and of bounded cohomology. The homotopy category of all monads over $\mathbf{U}$ will be denoted $\underline{\mathrm{Mon}}(\mathbf{U})$, its full subcategory of *linear monads* by $\underline{\mathrm{Mon}}^0(\mathbf{U})$.

Remark that, by definition, a (generalised) monad is just a special projective co-resolution, (5.6), whence (generalised) monads could as well be called *almost linear projective co-resolutions*.

It is clear, that being almost linear, or a monad, are properties of a complex which are stable under translation and which are preserved by forming mapping cones. Hence, almost linear complexes or monads form each a full triangulated subcategory of the homotopy category of "all" complexes.

Using the terms just defined, the proof of the Theorem above yields

122 COROLLARY 9.4.4. *With the notations as before, one has:*

(i) *There exists an exact equivalence from* $\mathrm{D^b}(\mathbf{U})$ *to* $\underline{\mathrm{Mon}}(\mathbf{U})$ *which associates to any object in* $\mathrm{D^b}(\mathbf{U})$ *an almost linear projective co-resolution.*
(ii) *The functor* $\mathbf{b}$ *establishes an exact equivalence from* $\mathrm{D^b}(\mathbf{U})$ *to* $\underline{\mathrm{Mon}}(\mathbf{UD})$*. It maps* $\mathrm{mod}(\mathbf{U})$ *isomorphically onto the abelian category* $\underline{\mathrm{Mon}}^0(\mathbf{UD})$ *of linear monads over* $\mathbf{UD}$.
(iii) *There exists a t-structure,* [**BBD82**, 1.3]*, on* $\mathrm{D^b}(\mathbf{U})$ *whose heart is given by all such complexes which admit a linear projective co-resolution.*

PROOF. Part (ii) follows immediately from the proof of (9.4.2). Namely, it has been observed there, that $\mathbf{d}$ (or $\mathbf{b}$) transforms any $\mathbf{U}$-module which is locally finite for the total degree into a linear complex. But if such a module is finitely generated over $\mathbf{U}$, the corresponding linear complex is necessarily bounded below, (9.2). That the cohomology of this complex is bounded (and also once again that the complex itself is bounded below) is then a formal consequence of the fact that this is true for the complexes associated to a generating family, say $\mathrm{Sym}(V)$ up to shifts as established directly during the proof of (9.4.2). This shows that $\mathbf{d}$ (or $\mathbf{b}$), if restricted to $\mathrm{D^b}(\mathbf{U})$ or $\mathrm{mod}(\mathbf{U})$, take the images claimed. As on the other hand $\underline{\mathrm{Mon}}(\mathbf{UD})$ is—by (6.2.1) (i) (b) and Verdier's criterion—a full subcategory of $\mathrm{D}(\mathrm{Mod}(\mathbf{UD}))$, which lies by definition "inside" $\mathrm{D^b}(\mathbf{UD})$, the theorem implies that $\underline{\mathrm{Mon}}(\mathbf{UD})$ is already equivalent to $\mathrm{D^b}(\mathbf{U})$ under either functor $\mathbf{d}$ or $\mathbf{b}$.

(i) is now obtained from (ii) by exchanging the roles of $\mathbf{U}$ and $\mathbf{UD}$: Given a complex X in $\mathrm{D^b}(\mathbf{U})$, represent its image under $\mathbf{b}$ by some finite complex in $\mathrm{D^b}(\mathbf{UD})$ and apply $\mathbf{b}^{-1}$. Making these choices "coherently"—without asking a logician—yields the exact equivalence. The t-structure claimed in (iii) is obtained from the natural one on $\mathrm{D^b}(\mathbf{UD})$ by applying $\mathbf{b}^{-1}$. □

REMARK. (b) As $\mathbf{U}$ is graded and strongly Gorenstein, it follows from (5.6) a priori that any object in $\mathrm{D^b}(\mathbf{U})$ admits a graded projective co-resolution. The additional information provided here is that such a projective co-resolution can be chosen to be almost linear!

Conversely, although the proof of (9.4.2) did not use explicitly the fact that $\mathbf{U}$ is strongly Gorenstein, one of the significant properties of such rings, namely the existence of projective co-resolutions—and then also automatically of complete resolutions, see (3.1) and (5.6)—follows from the (proof of the) theorem.

Now we can compile a dictionary between data over $\mathbf{U}(V,W)$ and the cor- **123**
responding data over $\mathbf{U}(\mathbf{D}(V,W))$ under the equivalences $\mathbf{b}_{(V,W)}$ for any linear super-space (V,W). The entries in the following list have to be read "up to isomorphism or natural equivalence", whatever applies:

(9.4.5)

	data over $\mathbf{U}=\mathbf{U}(V,W)$	$\overset{\mathbf{b}}{\underset{\longleftarrow}{\longrightarrow}}$ data over $\mathbf{UD}=\mathbf{U}(W^*,V^*)$
complexes	$\mathbf{K}_V$	$\mathbf{K}_{W^*}$
and	$\mathbf{U}$	$\det(V^*)=K(0,-v)$
modules	$\det(W)=K(0,-w)$	$\mathbf{UD}$
	$\bigwedge(W)$	$\bigwedge(V^*)$
gradings on $\mathrm{D}^{\mathrm{b}}(-)$	$[k](i,j)$	$[k+i+j](-j,-i)$
and the map on $\mathsf{G}_0(-)$	$\mathrm{cl}(\mathrm{Sym}(V))\cdot s^i\lambda^j$	$\mathrm{cl}(\mathrm{Sym}(W^*))\cdot(-1)^{v+w}s^{-w-j}\lambda^{-v-i}$
subcategories	$\mathrm{D}^{\mathrm{b}}_{\mathrm{perf}}(\mathbf{U})$	$\mathrm{D}^{\mathrm{b}}_{\mathrm{art}}(\mathbf{UD})$
of $\mathrm{D}^{\mathrm{b}}(-)$	$\mathrm{D}^{\mathrm{b}}_{\mathrm{art}}(\mathbf{U})$	$\mathrm{D}^{\mathrm{b}}_{\mathrm{perf}}(\mathbf{UD})$
	$\mathrm{D}^{\mathrm{b}}_{\mathrm{a/p}}(\mathbf{U})$	$\mathrm{D}^{\mathrm{b}}_{\mathrm{a/p}}(\mathbf{UD})$
quotient-categories	$\underline{\mathrm{MCM}}(\mathbf{U})$	$\mathrm{D}^{\mathrm{b}}(\mathrm{Proj}\text{-}\mathbf{UD})$
of $\mathrm{D}^{\mathrm{b}}(-)$	$\mathrm{D}^{\mathrm{b}}(\mathrm{Proj}\text{-}\mathbf{U})$	$\underline{\mathrm{MCM}}(\mathbf{UD})$
	$\underline{\mathrm{Prim}}(\mathbf{U})$	$\underline{\mathrm{Prim}}(\mathbf{UD})$
cohomological	$\mathrm{H}^k(-)_{(i,j)}$	$\mathrm{Ext}^{k+i+j}_{\mathbf{UD}}(K,\mathbf{b}(-))_{(-j,-i+v)}$
functors on $\mathrm{D}^{\mathrm{b}}(-)$	$\mathrm{Ext}^k_{\mathbf{U}}(-,\mathbf{U})_{(i,j)}$	$\mathrm{Ext}^{k+i+j}_{\mathbf{UD}}(\mathbf{b}(-),K)_{(-j,-i-v)}$

This is left to the reader. It follows immediately from the explicit description of $\mathbf{b}$ and (9.3).

Remark that $\mathbf{b}$ "reflects" the diagram (9.3.12) at the diagonal. Without even **124**
knowing $\mathbf{b}$ explicitly, one may deduce the above dictionary just from Beĭlinson's criterion and the following, easily established

LEMMA 9.4.6. *Let $F=(F^+,F^-)\colon (V,W)\to(V',W')$ be a linearly perfect morphism of linear super-spaces over K. Then*

(i) *F admits a unique factorization $F=f\circ\mathbf{D}(g)$, such that the even components of both f and g are identities.*
(ii) *If F^+ is an isomorphism, $\mathbf{U}(F)_*(\mathbf{K}_{V'})$ is isomorphic to $\mathbf{K}_V$.*
(iii) *If F^- is an isomorphism, then $\mathbf{U}(F)^!(\mathbf{K}_V)$ is isomorphic to $\mathbf{K}_{V'}$.*

PROOF. For (i) just observe that necessarily $f=(\mathrm{id}_{V'},F^-)$ and

$$g=(\mathrm{id}_{W^*},(F^+)^*).$$

The remaining assertions are once again easy exercises on Koszul complexes. $\square$

COROLLARY 9.4.7. *Let F be a linearly perfect morphism as above, $F=f\circ\mathbf{D}(g)$ its decomposition as in (9.4.6) ((i)). Then the following diagram of functors* *commutes—up to a natural isomorphism of functors:*

$$
\begin{array}{c|ccc}
 & \mathrm{D^b}(\mathbf{U}(V,W)) & \xrightarrow{\mathbf{b}_{(V,W)}} & \mathrm{D^b}(\mathbf{U}(\mathbf{D}(V,W))) \\
 & \Big\uparrow{\scriptstyle \mathbf{U}(\mathbf{D}(g))_*} & & \Big\uparrow{\scriptstyle \mathbf{U}(g)^*} \\
\mathbf{U}(F)_* & \mathrm{D^b}(\mathbf{U}(V',W)) & \xrightarrow{\mathbf{b}_{(V',W)}} & \mathrm{D^b}(\mathbf{U}(\mathbf{D}(V',W))) \\
 & \Big\uparrow{\scriptstyle \mathbf{U}(f)_*} & & \Big\uparrow{\scriptstyle \mathbf{U}(\mathbf{D}(f))^!} \\
 & \mathrm{D^b}(\mathbf{U}(V',W')) & \xrightarrow{\mathbf{b}_{(V',W')}} & \mathrm{D^b}(\mathbf{U}(\mathbf{D}(V',W'))) .
\end{array}
$$

PROOF. The commutativity of the upper square follows from the commutativity of the lower one by adjunction. For the lower square just use Beĭlinson's criterion, (9.4.2) and (9.4.6) (ii), (iii).[7] □

Now take for example $F = p_W : (V,W) \to (0,W)$, the projection onto the odd quotient of (V,W). Then the corollary implies immediately that $\mathbf{b}_{(V,W)}$ induces an exact equivalence between $\mathrm{D}^{\mathrm{b}}_{\mathrm{art}}(\mathbf{U})$ and $\mathrm{D}^{\mathrm{b}}_{\mathrm{perf}}(\mathbf{UD})$.

Similar arguments establish the other entries of (9.4.5).

For further use, let us also record the following which cannot be obtained directly from the Theorem 9.4.2 but rather from the explicit construction of $\mathbf{d}$ in its proof:

125 Let $F\colon (V,W) \to (V',W')$ be an *arbitrary* morphism of linear super-spaces. Then the forgetful functor—or "extension of scalars"—associated to $\mathbf{U}(F)$ is the identity on the underlying graded K-vectorspaces, hence preserves trivially the property of being locally finite for the total degree. As it is also exact, it passes "as such" to the derived categories, yielding an exact functor

$$\mathbf{U}(F)_*\colon \mathrm{D^b}(\mathrm{Mod}_{lf}(\mathbf{U}(V',W'))) \to \mathrm{D}(\mathrm{Mod}_{lf}(\mathbf{U}(V,W))) .$$

On the other hand, tensor product with $\mathbf{U}(\mathbf{D}(V,W))$ over $\mathbf{U}(\mathbf{D}(V',W'))$—or the "change-of-rings" along $\mathbf{U}(\mathbf{D}(F))$—transforms almost linear complexes over $\mathbf{U}(\mathbf{D}(V',W'))$ into such complexes over $\mathbf{U}(\mathbf{D}(V,W))$ and preserves homotopies. It induces consequently an exact functor $\mathbf{U}(\mathbf{D}(F))^*$ from the homotopy category of almost linear complexes over $\mathbf{U}(\mathbf{D}(V',W'))$ with its natural triangulated structure, (9.4.3), into the corresponding category of $\mathbf{U}(\mathbf{D}(V,W))$.

Now the explicit construction of $\mathbf{d}_{(V,W)}$ shows that one has a natural isomorphism of functors

$$\mathbf{d}_{(V,W)} \circ \mathbf{U}(F)_* = \mathbf{U}(\mathbf{D}(F))^* \circ \mathbf{d}_{(V',W')} , \tag{9.4.8}$$

and this implies (9.4.7) in case that F is linearly perfect.

Similarly, passing to the graded K-dual of a (complex of) graded module(s) is exact and transforms modules which are locally finite for the total degree into such, hence induces a functor—denote $(-)^\sharp$ as in (8.5.5)—from $\mathrm{D^b}(\mathrm{Mod}_{lf}(\mathbf{U}(V,W)))$ into $\mathrm{D^b}(\mathrm{Mod}_{lf}(\mathbf{U}(V,W)))^{\mathrm{op}}$.

Using then the appropriate sign convention for the dual of a complex—explained as for example in [**AGV72**, Exp. XVII, 1.1.5.1.]—it follows that there is a natural isomorphism of functors

$$\mathbf{d}^{\mathrm{op}}_{(V,W)}(-^\sharp) = \mathrm{Hom}_{\mathbf{U}(\mathbf{D}(V,W))}(\mathbf{d}_{(V,W)}(-), \mathbf{U}(\mathbf{D}(V,W))) \tag{9.4.9}$$

---

[7] To properly apply Beĭlinson's criterion one needs to first set up a natural transformation compatible with (9.4.6) (ii), (iii).

from $\mathrm{D}^{\mathrm{b}}(\mathrm{Mod}_{lf}(\mathbf{U}(V,W)))$ into the opposite of the homotopy category of almost linear complexes over $\mathbf{U}(\mathbf{D}(V,W))$.

Finally, if M and N are two graded $\mathbf{U}(V,W)$-modules such that their tensor product over K is locally finite for the total degree, one has the following isomorphism, natural in both arguments:[8]

$$\mathbf{d}(M \otimes_K N) = \mathbf{d}(M) \otimes_{\mathbf{U}(\mathbf{D}(V,W))} \mathbf{d}(N)\,. \tag{9.4.10}$$

Now we come back to our main theme and apply the correspondence to

126

9.5 Maximal Cohen–Macaulay modules and coherent sheaves. First, let us state explicitly the meaning of the functor $\mathbf{b}$ for maximal Cohen–Macaulay modules over a ring of the form $\mathrm{Sym}(V) \otimes_K \bigwedge(W)$:[9]

PROPOSITION 9.5.1. *The functor* $\mathbf{b}$ *induces an exact equivalence between the categories*

$$\underline{\mathrm{MCM}}(\mathbf{U}(V,W)) = \mathrm{D}^{\mathrm{b}}(\mathbf{U}(V,W))/\,\mathrm{D}^{\mathrm{b}}_{\mathrm{perf}}(\mathbf{U}(V,W))$$

and

$$\mathrm{D}^{\mathrm{b}}(\mathrm{Proj}\text{-}\mathbf{U}(\mathbf{D}(V,W))) = \mathrm{D}^{\mathrm{b}}(\mathbf{U}(\mathbf{D}(V,W)))/\,\mathrm{D}^{\mathrm{b}}_{\mathrm{art}}(\mathbf{U}(\mathbf{D}(V,W)))$$

with their respective natural triangulated structure. In particular:

(i) $\underline{\mathrm{MCM}}(\mathbf{U}(V,W))$ *comes equipped with a t-structure whose heart is equivalent to the abelian category* $\mathrm{Proj}\text{-}\mathbf{U}(W^*,V^*)$. *The associated cohomological functor, see* [**BBD82**, Thm. 1.3.6]*, is given by*

$$\mathrm{H}^0\colon \underline{\mathrm{MCM}}(\mathbf{U}(V,W)) \longrightarrow \mathrm{Proj}\text{-}\mathbf{U}(W^*,V^*)\,,$$
$$\mathrm{H}^0(M) = \mathbf{a}(\bigoplus_{i,j} \mathrm{Ext}^{i+j}_{\mathbf{U}(V,W)}(K,M)_{(-j,-i+w)})\,,$$

where the direct sum on the right is considered a (right) graded module over $\mathbf{U}(W^*,V^*) = \mathrm{Ext}_{\mathbf{U}(V,W)}(K,K)$ *with respect to the Yoneda product, cf.* ((9.3.2)).

(ii) *Under this equivalence, Tate cohomology over* $\mathbf{U}(V,W)$ *becomes transformed into coherent sheaf cohomology over* $\mathrm{Proj}\text{-}\mathbf{U}(W^*,V^*)$. *More precisely, for any graded maximal Cohen–Macaulay* $\mathbf{U}(V,W)$*-module* M*, (or any complex in* $\underline{\mathrm{D}}^{\mathrm{b}}(\mathbf{U}(V,W))$*), we have*

$$\begin{aligned}\widehat{\mathrm{H}}^i_{\mathbf{U}(V,W)}(M \otimes_K (\det W)^{-1}) &= \underline{\mathrm{Ext}}^i_{\mathbf{U}(V,W)}(K, M \otimes_K (\det W)^{-1})\\ &= \mathrm{Ext}^i_{\mathrm{Proj}\text{-}\mathbf{U}(W^*,V^*)}(\mathbf{a}(\mathbf{U}(W^*,V^*)), \mathbf{a}\circ\mathbf{b}(M))\\ &=: \mathrm{H}^i(\mathrm{Proj}\text{-}\mathbf{U}(W^*,V^*), \mathbf{a}\circ\mathbf{b}(M))\,.\end{aligned}$$

(iii) *The duality for Tate cohomology over* $\mathbf{U}(V,W)$*—see* (7.7.5) *and* (8.5.5) *&* (8.5.6)*—translates into Grothendieck–Serre duality for the cohomology of (complexes of) coherent sheaves of modules on the ringed space* $\mathrm{Proj}\text{-}\mathbf{U}(W^*,V^*)$*—formulated in* [**OP84**] *for the analogous* $\mathbb{Z}/2\mathbb{Z}$*-graded case.*

PROOF. This consists just in a reformulation of (part of) the dictionary (9.4.5) and we restrict ourselves to explain (iii). (The fixed linear super-space (V,W) is again dropped from the notation.)

---

[8] This remarkable formula deserves more attention; it says the BGG equivalence is one of tensor triangulated categories. See Remark A.9.2 for more details.

[9] Here and below $w = \dim_K W$ and $v = \dim_K V$.

127 From Remark (a) in (9.2) recall that the canonical module of **UD** over K is given by $\omega_{\mathbf{UD}} = \mathbf{UD}\otimes_K \det(W^*,V^*) \cong \mathbf{UD}(-w,v)$. The table (9.4.5) shows then that one has an isomorphism in $\mathrm{D^b}(\mathbf{UD})$:

$$\omega_{\mathbf{UD}} \xrightarrow{\cong} \mathbf{b}((\det W)[v-w](-v,w)) = \mathbf{b}((\det V)[v-w])\,.$$

Let $\omega = \mathbf{a}(\omega_{\mathbf{UD}})$ denote the sheafification of the dualizing module. Not surprisingly, it will serve as the dualizing sheaf on Proj-**UD**, and the induced sheaf on the "underlying" projective super-space $\mathbb{P}(W^*,V^*)$—see (9.2) and Remark (c) after ((9.3.6))—is just the "Berezinian sheaf", [**Lei83**], [**OP84**].

If now F is any complex in $\mathrm{D^b}(\text{Proj-}\mathbf{UD})$, M any MCM over **U**, (or, as well, any complex in $\mathrm{D^b}(\mathbf{U})$), such that F is isomorphic to the sheafification of $\mathbf{b}(M)$, we get from (ii) above isomophisms:

$$\mathrm{Ext}^{w-1}_{\text{Proj-}\mathbf{UD}}(F,\omega) \cong \underline{\mathrm{Ext}}^{v-1}_{\mathbf{U}}(M,\det V)$$

and

$$\mathrm{H}^0(\text{Proj-}\mathbf{UD},F) \cong \widehat{\mathrm{H}}^0(M\otimes_K(\det W)^{-1})\,.$$

Recall, (7.7.5) (iii) & (iv) and (8.5.5) & 8.5.6, that the duality for Tate cohomology over **U** is obtained by combining the Yoneda product with the trace τ_K, see also (8.5.7) (v), which hence yields in view of the just exhibited isomorphisms of cohomology groups the diagram

$$\begin{array}{ccccc}
\underline{\mathrm{Ext}}^{v-1}_{\mathbf{U}}(M,\det V)\times \widehat{\mathrm{H}}^0(M\otimes_K(\det W)^{-1}) & \xrightarrow{-\circ-} & \det(V,W)\otimes_K \widehat{\mathrm{H}}^{v-1} & \xrightarrow{\tau_K} & K\\
\updownarrow\cong \qquad\qquad \updownarrow\cong & & \updownarrow\cong & & \|\\
\mathrm{Ext}^{w-1}_{\text{Proj-}\mathbf{UD}}(F,\omega)\times \mathrm{H}^0(\text{Proj-}\mathbf{UD},F) & \xrightarrow{-\circ-} & \mathrm{H}^{w-1}(\text{Proj-}\mathbf{UD},\omega) & \longrightarrow & K\,,
\end{array}$$

in which the upper row describes the duality for Tate cohomology over **U** whereas the lower row is the expected form of Grothendieck–Serre duality on Proj-**UD**. Expliciting the trace on $\mathrm{H}^{w-1}(\text{Proj-}\mathbf{UD},\omega)$ in terms of "super-differential" forms, it becomes then justified—as mentioned—to interpret τ_K as "the integration of differential forms over the fundamental cycle of Proj-**UD**". □

REMARK. (a) If $i_{W^*}\colon (W^*,0)\to(W^*,V^*)$ denotes as usual the inclusion of the even subspace of (W^*,V^*), (9.1) (iii), and—abusively, as in ((9.3.7))—$(i_{W^*})^*$ the "inverse image functor" associated to the morphism of ringed spaces from Proj-$\mathbf{U}(W^*,V^*)$ to Proj-$\mathbf{U}(W^*,0)$ induced by i_{W^*}, then one has obviously that
128 $\mathbf{a}(\mathbf{U}(W^*,V^*))$, the structure sheaf of Proj-$\mathbf{U}(W^*,V^*)$, is the inverse image $(i_{W^*})^*(\mathcal{O}_{\mathbb{P}(W^*)})$ of the usual structure sheaf on the projective space $\mathbb{P}(W^*)$.

Hence, there is an equality of cohomological functors

$$\mathrm{H}^i(\text{Proj-}\mathbf{U}(W^*,V^*),-) = \mathrm{H}^i(\mathbb{P}(W^*),(i_{W^*})_*(-))\,,$$

so that indeed Tate cohomology over $\mathbf{U}(V,W)$ "is" nothing but coherent sheaf cohomology on $\mathbb{P}(W^*)$, the projective space of all "odd points" of the linear super-space (V,W), (9.1) (ii).

Given the natural duality between geometric and odd points, this indicates once again in which sense Tate cohomology "complements" sheaf cohomology on linear super-spaces.

As a further "application", remark that sheaf cohomology together with Tate cohomology just detect those complexes over $\mathbf{U}(V,W)$ which are at the same time perfect and of artinian cohomology. Now, considering the geometric points only, hence forgetting the structure over the exterior algebra, the "Lemme d'acyclicité" implies that any perfect representative of such a complex must have at least $\dim_K V$ many non-zero terms. Dually, that is considering the odd points or applying the same argument to the image under $\mathbf{b}$, the (total) length of the cohomology of such a complex is bounded below by $\dim_K W$. This kind of argument can be found for example in the work of G. Carlsson, [**Car86**], where this dictionary is used—although not in the same language—to bound the cohomology of spaces admitting free actions of elementary abelian 2-groups.

Coming back to the interpretation of Tate cohomology on projective space, it should be an interesting exercise to translate the fundamental theorems on this cohomology—like Serre's Theorems A and B or the Theorem of Riemann–Roch—into statements about Tate cohomology of modules over rings of the form $\mathrm{Sym}(V)\otimes_K \bigwedge(W)$.

Here, we will instead restrict ourselves to just emphasise the consequences of the existence of a (natural) t-structure on $\underline{\mathrm{MCM}}(\mathbf{U}(V,W))$ for any linear super-space (V,W).

To begin with, this allows one to define higher algebraic K-groups for MCM's over $\mathbf{U}(V,W)$ and these groups may serve as "stabilized" K-groups for $\mathbf{U}(V,W)$ as asked for in (4.9):

DEFINITION AND PROPOSITION 9.5.2. *Let* D *be any t-category with heart* C. **129**
Then define the higher algebraic K-groups $\mathsf{K}_i(\mathsf{D})$ *to be the K-groups* $\mathsf{K}_i(\mathsf{C})$ *as defined by D. Quillen,* [**Qui73**].

For any linear super-space (V,W) *set* $\underline{\mathsf{K}}_i(\mathbf{U}(V,W)) := \mathsf{K}_i(\underline{\mathrm{MCM}}(\mathbf{U}(V,W)))$, *the category of maximal Cohen–Macaulay modules over* $\mathbf{U}(V,W)$ *modulo projectives endowed with the t-structure from* (9.5.1) ((i)).

Then one has

(i) $\mathsf{K}_i(\mathbf{U}(V,W)) = \mathsf{G}_i(\mathbf{U}(V,W)) = \mathsf{K}_i(K)\otimes_{\mathbb{Z}}\mathbb{Z}[s,s^{-1},\lambda,\lambda^{-1}]$,

(ii) *there are short exact sequences of abelian groups*

$$0 \longrightarrow \mathsf{K}_i(\mathbf{U}(V,W)) \xrightarrow{(1-\lambda^{-1})^w} \mathsf{G}_i(\mathbf{U}(V,W)) \longrightarrow \underline{\mathsf{K}}_i(\mathbf{U}(V,W)) \longrightarrow 0$$

and, consequently,

(iii) $\underline{\mathsf{K}}_i(\mathbf{U}(V,W)) = \mathsf{K}_i(K)\otimes_{\mathbb{Z}}\mathbb{Z}[s,s^{-1},g]/(g^w)$, *where* g *represents the class of* $(1-\lambda^{-1})$ *as in* ((9.3.8)). *This K-group is naturally isomorphic to* $\mathsf{K}_i(\mathbb{P}(W^*))[\lambda,\lambda^{-1}]$, *the isomorphism induced by the functor* $\mathbf{b}_{(V,W)}$.

PROOF. In view of the Bernšteĭn–Gelfand–Gelfand correspondence and the dictionary deduced from it, the assertions above just come down to the calculation of higher K-groups of projective space as done by D. Quillen in [**Qui73**, Prop. 4.3]. For the last statement in (iii), remark that the direct image functor $(i_{W^*})_*$ is t-exact with respect to the natural t-structures on $\mathrm{D}^{\mathrm{b}}(\mathrm{Proj}\text{-}\mathbf{U}(W^*,V^*))$ and $\mathrm{D}^{\mathrm{b}}(\mathrm{Proj}\text{-}\mathbf{U}(W^*,0))$ and that it admits $(i_{W^*})^*$ as an exact left adjoint. Combining now the foregoing with [**BBD82**, 1.3.7. (iii)] and [**Qui73**, §3], it follows the existence of natural isomorphisms of abelian groups

$$\underline{\mathsf{K}}_i(\mathbf{U}(V,W)) \xrightarrow{\mathsf{K}_i(\mathbf{b})} \mathsf{K}_i(\mathrm{Proj}\text{-}\mathbf{U}(W^*,V^*)) \xrightarrow{\mathsf{K}_i((i_{W^*})_*)} \mathsf{K}_i(\mathrm{Proj}\text{-}\mathbf{U}(W^*,0))$$

and

$$\mathsf{K}_i(\mathbb{P}(W^*))\otimes_{\mathbb{Z}}\mathbb{Z}[\lambda,\lambda^{-1}] \longrightarrow \mathsf{K}_i(\mathrm{Proj}\text{-}\mathbf{U}(W^*,0))\,.$$

□

REMARKS. (b) Analogously, one may define in a natural way higher K-groups for the category $\underline{\mathrm{Prim}}(\mathbf{U}(V,W))$ of primitive objects, ((9.3.8)), and show that these groups are canonically isomorphic to the corresponding groups for $\mathbb{P}(V) \times \mathbb{P}(W^*)$, the scheme of all points of the linear super-space (V, W).

(c) Above, we have reduced the (definition and) calculation of K-groups for MCM's to the one for projective spaces by means of the Bernšteĭn–Gelfand–Gelfand corre-
130 spondence **b**. In fact, analysing D. Quillen's result, or, even more instructively, its extension by R. G. Swan to encompass the case of non-singular quadrics [**Swa68**], one observes that this correspondence is used there—implicitly—and that the determination of the K-groups of the projective scheme in question is deduced from the explicit knowledge of "all" MCM's over the Yoneda-Ext-algebra of the residue class field of the associated homogeneous coordinate ring.

(In case of projective space $\mathbb{P}(V)$, this Ext-algebra is $\bigwedge(V^*)$, see ((9.3.2)) (iv), and the category $\underline{\mathrm{MCM}}(\bigwedge(V^*))$ is equivalent to $\underline{\mathrm{mod}}(\bigwedge(V^*))$ by (8.1). For any non-singular quadric, the Ext-algebra is the homogenised Clifford-algebra of the associated quadratic form, and a module over it is MCM iff it is finitely generated and the homogenising parameter is a non-zero-divisor in it. For more details along these lines, see also the appendix of [**BEH87**].)

Next, we will investigate a little bit closer the t-structure on the category of maximal Cohen–Macaulay $\mathbf{U}(V,W)$-modules modulo projectives by using only local information at the odd points of (V, W).

This will result in a generalisation of the "monadic" description of vector bundles on usual projective space as given in [**BGG78**] to yield a representation of coherent sheaves of modules in terms of monads satisfying certain "perversity-conditions". Finally, we will give a "normal form" for such a monad, making precise the statement that "any coherent sheaf of modules on projective space is entirely determined—up to isomorphism—by a finite number of cohomology groups".

To begin with, we have to recall some elementary facts on the natural t-structure on $\mathrm{D^b}(\mathrm{Proj}\text{-}\mathbf{U}(V,W))$—or just on $\mathrm{D^b}(\mathbb{P}(V))$.

As done already in case of i_V, for any morphism of linear super-spaces $F\colon (V,W) \to (V',W')$—with F^+ surjective—we will denote by

$$F_*\colon \mathrm{D^b}(\mathrm{Proj}\text{-}\mathbf{U}(V',W')) \to \mathrm{D^b}(\mathrm{Proj}\text{-}\mathbf{U}(V,W))$$

the direct image functor associated to the morphism $\mathrm{Proj}(F)$ of ringed spaces. F^* or $F^!$ will denote its left and right adjoint—if either exists.

We introduce further the following

NOTATIONS 9.5.3. Let (V, W) be a linear super-space over K, λ a non-zero linear form on V, corresponding to a geometric points $\langle\lambda\rangle$ of (V,W) in $\mathbb{P}(V)$. Let $K(\langle\lambda\rangle)$ denote the residue class field at that point of $\mathbb{P}(V)$.

Then we set for any complex C in $\mathrm{D^b}(\mathrm{Proj}\text{-}\mathbb{P}(V,W))$:

$$T_i(\langle\lambda\rangle; C) := \mathrm{H}^{-i}(\lambda^* \circ (i_V)_*(C)) = \mathrm{Tor}_i^{\mathcal{O}_{\mathbb{P}(V)}}((i_V)_*(C), K(\langle\lambda\rangle))$$

131 and

$$S_i(C) := \{p \in \mathrm{Proj}\text{-}\mathrm{Sym}(V)\colon\ \mathrm{Tor}_i^{\mathcal{O}_{\mathbb{P}(V)}}((i_V)_*(C), K(p)) \neq 0\}\,.$$

Here it is to be understood that we have summed up over the still existing "exterior degrees" for $(i_V)_*(C)$, so that the "correct" definition of $T_i(\langle\lambda\rangle; C)$ should read $\bigoplus_j \mathrm{H}^{-i}(\lambda^* \circ (i_V)_*(C(0,j)))$.

The functors $T_i(\langle\lambda\rangle;-)$ are easily described in terms of the equivalence $\mathbf{b}$: Let M be a maximal Cohen–Macaulay module over $\mathbf{U}(W^*,V^*)$, $z\colon K\to V^*$ the odd point of (W^*,V^*) which corresponds to λ by duality. Then (9.4.7) shows that one has

$$\bigoplus_j \mathrm{H}^{-i}(\lambda^*\circ(i_V)_*(\mathbf{b}(M)(0,k)))=\bigoplus_j \widehat{\mathrm{H}}^{-i+j}(z_*\circ\mathbf{U}(p_{V^*})^!(M(-j,0)))\,,$$

where $p_{V^*}=\mathbf{D}(i_V)\colon(W^*,V^*)\to(0,V^*)$ is the projection onto the odd quotient space of $\mathbf{D}(V,W)$.

The right hand side can be evaluated explicitly. If $K\langle z\rangle$ denotes the sub-algebra of $\bigwedge(V^*)$ generated by the image of z, the cohomology group in question is just

$$\bigoplus_j \mathrm{H}^{-i+j}(\mathrm{Hom}_{K\langle z\rangle}(\mathbf{CR}_{K\langle z\rangle}(K),z_*\,\mathrm{RHom}_{\mathbf{U}(W^*,V^*)}(\textstyle\bigwedge(V^*),M(-j,0))))\,,$$

which looks worse than it is:

- A complete resolution of K over $K\langle z\rangle$ is given by the periodic complex
$$\cdots\xrightarrow{z}K\langle z\rangle(0,k)\xrightarrow{z}K\langle z\rangle(0,k+1)\xrightarrow{z}\cdots\,,$$
(where z is also identified with $z(1)$),
- $\bigwedge(V^*)$ is resolved over $\mathbf{U}(W^*,V^*)$ by the Koszul complex
$$(\textstyle\bigwedge^k(W^*)\otimes_K\mathrm{Sym}(W^*)\otimes_K\bigwedge(V^*)[k](-k,0),\,d_{W^*}\otimes 1)\,,$$
where d_{W^*} represents the ordinary Koszul differential associated to W^*.

As a simple example, which appears already in [**BGG78**], assume $W=0$—in which case one may forget about i_V and the Koszul complex above—so that **132**
$M=\bigoplus M_i^j$ is just a finite direct sum (indexed by i) of graded $\bigwedge(V^*)$-modules. Identifying z—or rather $z(1)$—with a 1-form in $\bigwedge(V^*)$, multiplication with it turns M into a complex of vectorspaces

$$\cdots\xrightarrow{z}M^j\xrightarrow{z}M^{j+1}\xrightarrow{z}\cdots$$

whose cohomology is the desired object.

The behavior of the homological functors $T_i(\langle-\rangle;-)$ is well-known:

LEMMA 9.5.4. *Let V be a K-vectorspace of dimension v, x any point in $\mathbb{P}(V)$ and F a coherent sheaf of $\mathcal{O}_{\mathbb{P}(V)}$-modules. Denote F_x the localization of F at x. Then the following assertions are equivalent for any non-negative i:*

(i) $T_i(x;F)\neq 0$
(ii) $\mathrm{projdim}_{\mathcal{O}_{\mathbb{P}(V),x}}F_x\geq i$
(iii) $\mathrm{depth}_{\mathcal{O}_{\mathbb{P}(V),x}}F_x\leq v-1-i$
(iv) $\mathrm{H}^j_{\{x\}}(F_x)\neq 0$ *for some* $j\leq v-1-i$
(v) $\bigoplus_{j=0}^{v-1-i}\mathrm{Ext}^{v-1-j}_{\mathcal{O}_{\mathbb{P}(V),x}}(F_x,\omega_{\mathbb{P}(V),x})\neq 0$
(vi) *x is in the support of* $\bigoplus_{j=0}^{v-1-i}\mathcal{E}xt^{v-1-j}_{\mathcal{O}_{\mathbb{P}(V)}}(F,\omega_{\mathbb{P}(V)})$.[10]

PROOF. (i) $\Leftrightarrow$ (ii) is the existence of a minimal projective resolution of F_x over the local ring $\mathcal{O}_{\mathbb{P}(V),x}$.

[10] This is the sheaf Ext.

(ii) $\Leftrightarrow$ (iii) is the theorem of Auslander–Buchsbaum–Serre: $\mathcal{O}_{\mathbb{P}(V),x}$ is a regular local ring of dimension $v-1$ and hence

$$\operatorname{projdim}_{\mathcal{O}_{\mathbb{P}(V),x}} F_x + \operatorname{depth}_{\mathcal{O}_{\mathbb{P}(V),x}} F_x = v-1$$

(iii) $\Leftrightarrow$ (iv) is the characterization of depth by local cohomology.

(iv) $\Leftrightarrow$ (v) is Grothendieck–Serre duality: The K-dual of $\mathrm{H}^j_{\{x\}}(F_x)$ is isomorphic to $\operatorname{Ext}^{v-1-j}_{\mathcal{O}_{\mathbb{P}(V),x}}(F_x, \omega_{\mathbb{P}(V),x})$.

(v) $\Leftrightarrow$ (vi) is the fact that localization and forming Ext's commutes. □

COROLLARY 9.5.5. *With notations as before,*

(i) *$S_i(F)$ contains $S_{i+1}(F)$ and $S_0(F)$ equals the support of F.*
(ii) *Any $S_i(F)$ is a closed subset of $\mathbb{P}(V)$.*
(iii) *The codimension of $S_i(F)$ in $\mathbb{P}(V)$ is at least i.*
(iv) *F satisfies Serre's condition S_i, iff $S_{v-i}(F) = \varnothing$.*
(v) *If $\dim F \leq r$, then $S_{v-1-r}(F) = \cdots = S_0(F) = \operatorname{Supp}(F)$.*

133 *In other words, if we consider the function $c_F(i) = \operatorname{codim}_{\mathbb{P}(V)}(S_i(F))$, it has the following properties:*

(i') *It is non-decreasing and its "initial value" $c_F(0)$ equals the codimension of the support of F by* (i).
(ii') *It stays above or on the diagonal by* (iii).
(iii') *It hits the diagonal at least once, namely for $i - v - 1 - \dim F$, by* (v). □

These properties are enough to prove

PROPOSITION 9.5.6. *Let C be a complex in $\mathrm{D}^{\mathrm{b}}(\operatorname{Proj}\text{-}\mathbf{U}(V,W))$. Then C is quasi-isomorphic to $\mathrm{H}^0(C)$, that is, there is no other cohomology than possibly in degree zero or, equivalently, C belongs to the heart of the natural t-structure on $\mathrm{D}^{\mathrm{b}}(\operatorname{Proj}\text{-}\mathbf{U}(V,W))$, if and only if*

(i) *$S_i(C) = \varnothing$ for $i < 0$ and*
(ii) *$\operatorname{codim}_{\mathbb{P}(V)} S_i(C) \geq i$ for $i \geq 0$.*

PROOF. It follows from the foregoing that the two conditions are necessary. To prove that they are sufficient, we show that with

- $i(C) := \min(i : \mathrm{H}^i(C) \neq 0)$ and $m(C) := \max(i : \mathrm{H}^i(C) \neq 0)$,
- $r := \dim \operatorname{Supp} \mathrm{H}^{i(C)}(C)$,

one has:

(a) $S_{-m(C)}(C) \neq \varnothing$ and
(b) $\operatorname{codim}_{\mathbb{P}(V)} S_{v-1-r-i(C)}(C) \leq v-1-r$,

which evidently will suffice.

(a) and (b) both follow from an examination of the spectral sequence

$$E_2^{i,j}(x;C) = T_{-i}(x; \mathrm{H}^j(C)) \Longrightarrow T_{-i-j}(x;C)\,,$$

x any point in $\mathbb{P}(V)$.

By definition of $m(C)$, for any x in $\mathbb{P}(V)$ the terms $E_2^{-i,j}(x;C)$ vanish if either

- $j - i \geq m(C) + 1$ or
- $j - 1 = m(C)$, but $i \neq 0$, or
- $j - 1 = m(C) - 1$ and $i \neq 0, 1$.

Hence, $T_{-m(C)}(x;C) = E_2^{0,m(C)}(x;C) = T_0(x;\mathrm{H}^{m(C)}(C))$ which proves (a).

To establish (b), remark first that by (9.5.5) (v) one has

$$S_{v-1-r}(\mathrm{H}^{i(C)}(C)) = \mathrm{Supp}(\mathrm{H}^{i(C)}(C))\,,$$

which by assumption is of dimension r. As on the other hand for any $k>0$ one obtains from (9.5.5) (iii) that

$$\mathrm{codim}_{\mathbb{P}(V)} S_{v-1-r+k+1}(\mathrm{H}^{i(C)+k}(C)) \geq v-1-r+k+1 > v-1-r\,,$$ **134**

the above spectral sequences imply that $S_{v-1-t-i(C)}(C)$ contains all r-dimensional components of $\mathrm{Supp}(\mathrm{H}^{i(C)}(C))$—whence (b). □

To simplify our next statement, we translate as above the notions $T_i(\langle\lambda\rangle;-)$ and $S_i(-)$ into Tate cohomology:

DEFINITION 9.5.7. Let $w\colon K\to W$ be a non-zero odd point of the linear super-space (V,W). Then we set

$$\widehat{\mathrm{H}}^i_w(M) := \bigoplus_j \widehat{\mathrm{H}}^{i+j}(w_*\circ \mathbf{U}(p_W)^!(M(-j,0)))$$

for any object M in $\mathrm{D^b}(\mathbf{U}(V,W))$.

The scheme-theoretic support of this family of functors $\widehat{\mathrm{H}}^i_w(-)$ in $\mathbb{P}(W^*)$ will then be denoted

$$\widehat{S}_i(M) := \mathrm{Supp}_{\mathbb{P}(W^*)}\,\widehat{\mathrm{H}}^{-i}_{(-)}(M)\,.$$

More precisely, it contains all those homogeneous primes $\mathfrak{q}$ in Proj-Sym(W^*), for which there exists a finite field extension L of K, and an L-valued odd point $w_L\colon L\to W\otimes_K L$, such that the corresponding point in $\mathbb{P}(W^*\otimes_K L)$ lies over $\mathfrak{q}$ and such that $\widehat{\mathrm{H}}^{-i}_{w_L}(M\otimes_K L)\neq 0$.

Just by construction, one has

$$\widehat{\mathrm{H}}^i_w(M) = T_{-i}(\langle \mathbf{D}(w)\rangle;\mathbf{ab}(M)) \quad\text{and}\quad \widehat{S}_{-i}(M) = S_{-i}(\mathbf{ab}(M))\,.$$

Hence we get the following description of the t-structure on MCM's:

THEOREM 9.5.8. *Let (V,W) be a linear super-space over K, M any complex of modules in $\mathrm{D^b}(\mathbf{U}(V,W))$. Then the following are equivalent:*

(i) *The image of M in $\underline{\mathrm{D}}^{\mathrm{b}}(\mathbf{U}(V,W))$ is in the heart of the natural t-structure.*
(ii) *M admits an essentially linear projective co-resolution.*
(iii) *$\mathbf{ab}(M)$ has only cohomology in (complex-)degree zero.*
(iv) *$\widehat{S}_i(M)=\varnothing$ for $i<0$ and $\mathrm{codim}_{\mathbb{P}(W^*)}\widehat{S}_i(M)\geq i$ for $i\geq 0$.*

PROOF. (i) ⇔ (iii) is the definition, (9.5.1) (i).

(ii) ⇔ (iii) as by the properties of the correspondence $\mathbf{b}$, to say that the sheafification of $\mathbf{b}(M)$ has only cohomology in (complex-)degree zero just means that up to a perfect complex, M lies in the pre-image of $\mathrm{mod}(\mathbf{U}(\mathbf{D}(V,W)))$. But that signifies, by (9.4.4), that M admits a projective co-resolution which becomes linear after a finite number of terms.

(iii) ⇔ (iv) just translates the characterization of the natural t-structure on **135**
$\mathrm{D^b}(\mathrm{Proj}\text{-}\mathbf{U}(W^*,V^*))$, which was obtained in (9.5.6), by means of $\mathbf{b}$. □

To complete further our dictionary, let us point out the significance of the existence of MCM-approximations over $\mathbf{U}(V,W)$ for the study of complexes of coherent sheaves of modules over $\operatorname{Proj}\text{-}\mathbf{U}(\mathbf{D}(V,W))$.

Recall that, given a complex X in $\mathrm{D^b}(\mathbf{U}(V,W))$, an MCM-approximation $\mathbf{M}(X)$ is by definition, (5.6), a maximal Cohen–Macaulay $\mathbf{U}(V,W)$-module which, considered as a complex, is quasi-isomorphic to X up to a perfect complex.

Now a complex C in $\mathrm{D^b}(\mathbf{U}(V,W))$ is isomorphic to a single MCM if and only if it satisfies

(i) $\operatorname{Ext}^k_{\mathbf{U}(V,W)}(\mathbf{U}(V,W),C(i,j))=0$ except for $k=0$, i.e. the cohomology of C is concentrated in (complex-)degree zero, and

(ii) $\operatorname{Ext}^k_{\mathbf{U}(V,W)}(C(i,j),\mathbf{U}(V,W))=0$ except for $k=0$, which means that the only cohomology module which may exist by (i) is MCM.

Applying our dictionary, this translates into the following conditions for $\mathbf{b}(C)$:

(i') $\mathbf{b}(C)$ admits a *linear projective co-resolution*, or, equivalently, is isomorphic to a linear monad in $\mathrm{D}(\mathbf{U}(W^*,V^*))$, by (9.4.4) (ii), and

(ii') $\operatorname{Ext}^{k+i+j}_{\mathbf{U}(W^*,V^*)}(\mathbf{b}(C)(-j,-i),\det(V^*)) = 0$ for any $k\neq 0$, or, equivalently: $\mathbf{b}(C)\otimes_K(\det V^*)^{-1}$ admits a *linear projective resolution.*

Remember furthermore, that MCM-approximations are unique up to projective summands, hence over $\mathbf{U}(V,W)$ up to direct sums of graded free modules of the form $\mathbf{U}(V,W)(i,j)$. This says that we may add to $\mathbf{b}(C)$ complexes of the form

$$\mathbf{b}(\mathbf{U}(V,W)(i,j))=(\det V^*)[i+j](-j,-i)=K[i+j](-j,-i-v)\,,$$

or, connecting (i') and (ii') to a single complete resolution of $\mathbf{b}(C)$, the resulting acyclic complex of finitely generated graded free modules determined by C is modified by direct summands which constitute shifted complete resolutions of K itself. Remark that "the" complete resolution of K was determined explicitly in (8.5.7). Hence, translating the existence of MCM-approximations yields

136 PROPOSITION 9.5.9. *Let Y be any complex in $\mathrm{D^b}(\operatorname{Proj}\text{-}\mathbf{U}(W^*,V^*))$. Then there exits:*

(i) *A complex of graded $\mathbf{U}(W^*,V^*)$-modules $\mathbf{C}(Y)$, bounded below and with bounded cohomology, whose term in (complex-)degree k is a finite direct sum of modules of the form $\mathbf{U}(W^*,V^*)(i,j)$ with $i+j=k$.*

(ii) *A complex of graded $\mathbf{U}(W^*,V^*)$-modules $\mathbf{P}(Y)$, bounded above and with bounded cohomology, whose term in (complex-)degree k is a finite direct sum of modules of the form $\mathbf{U}(W^*,V^*)(i,j)$ with $i+j=k-v$.*

(iii) *A morphism of complexes $\mathbf{N}(Y)\colon\mathbf{P}(Y)\to\mathbf{C}(Y)$ which induces an isomorphism of cohomology and such that the sheafification of the image of $\mathbf{N}(Y)$ is a (necessarily finite) complex isomorphic to Y in $\mathrm{D^b}(\operatorname{Proj}\text{-}\mathbf{U}(W^*,V^*))$.*

(iv) *The morphism $\mathbf{N}(Y)$ is unique up to isomorphism if one imposes the "minimality-condition" that the mapping cone of $\mathbf{N}(Y)$ does not contain a shifted complete resolution of K as a direct summand.*

One may hence call the morphism $\mathbf{N}(Y)\colon\mathbf{P}(Y)\to\mathbf{C}(Y)$ the *minimal monadic representation* of the given complex of sheaves Y.

PROOF. This is immediate from the foregoing: If Y' denotes any representative of Y in $\mathrm{D^b}(\mathbf{U}(W^*,V^*))$, set $X'=\mathbf{b}_{(W^*,V^*)}(Y')$ and $\mathbf{M}(X')$ its unique MCM-approximation over $\mathbf{U}(V,W)$ which does not contain any free summand. Then

$\mathbf{C}(Y)$ is the linear monad associated to $\mathbf{b}_{(V,W)}(\mathbf{M}(X'))$, by (i') above, $\mathbf{P}(Y)$ its minimal projective resolution, which has the required properties in view of (ii') above. $\mathbf{N}(Y)$ is then the natural map, whose image is by construction quasi-isomorphic to $\mathbf{b}_{(V,W)}(\mathbf{M}(X'))$ and differs from $Y' = \mathbf{b}_{(V,W)}(X')$ only by a complex which has artinian cohomology, hence its sheafification is isomorphic to Y again.

Starting with another lifting Y'' of Y, it differs from Y' only by a complex with artinian cohomology, hence the corresponding objects X' and X'' differ only by a perfect complex, whence they have the same minimal MCM-approximation. This shows the uniqueness of the construction. □

REMARK. (d) If $v = \dim_K V \neq 1$, the (isomorphism class of the) complex Y (in $\mathrm{D^b}(\mathrm{Proj}\text{-}\mathbf{U}(W^*, V^*))$) can be recovered from the (isomorphism class of the) mapping cone over $\mathbf{N}(Y)$ (in the category of complexes): By construction, we can then distinguish the linear subcomplex $\mathbf{C}(Y)$ and the corresponding quotient complex $\mathbf{P}(Y)[1]$ just from the occurring degrees.

Let us emphasise that we deal indeed with *isomorphism classes* of complexes— **137**
and not with *homotopy classes*. The reader is asked to supplement (9.5.9) by convincing themself that in fact the homotopy classes of the occurring complete resolutions are in bijection with the isomorphism classes of objects in the (equivalent) categories of primitive objects $\underline{\mathrm{Prim}}(\mathbf{U}(V, W))$ or $\underline{\mathrm{Prim}}(\mathbf{U}(W^*, V^*))$.

As an example of the foregoing—and to justify the title of this chapter—let us mention the

9.6 Applications to projective geometry.

(9.6.1) Considering the last result first, assume $V = 0$ in (9.5.9). Then $\mathbf{U}(W^*, V^*) = \mathrm{Sym}(W^*)$, the polynomial ring over W^*. For any object Y in $\mathrm{D^b}(\mathbb{P}(W^*))$, by construction both $\mathbf{P}(Y)$ and $\mathbf{C}(Y)$ are linear complexes, which are furthermore finite as $\mathrm{Sym}(W^*)$ is of finite global dimension. It follows that $\mathbf{N}(Y)$ is necessarily an isomorphism and the minimality condition just ensures that neither $\mathbf{P}(Y)$ or $\mathbf{C}(Y)$ contain a shifted Koszul complex as a direct summand. In particular, the mapping cone over such a morphism $\mathbf{N}(Y)$ is certainly contractible—in accordance with the fact that the category of primitive objects over a polynomial ring is trivial.

To put it differently, if we consider in general a linear super-space (V, W) in which either the even or odd summand is zero, the diagram in (9.3.12) degenerates, leaving essentially a unique row or column, and the "minimal monadic representation" of (9.5.9) is already completely determined by either $\mathbf{P}(Y)$ or $\mathbf{C}(Y)$.

Now let us summarise the various descriptions of complexes of coherent sheaves over a projective space $\mathbb{P}(W^*)$ which are contained in the above—and which all already appeared in the literature, at least in some disguise:

(9.6.2) Assume given a finite dimensional vectorspace W over some field K. Then the derived category $\mathrm{D^b}(\mathbb{P}(W^*))$ of complexes of sheaves of modules on the projective space $\mathbb{P}(W^*)$ whose non-zero cohomology sheaves are finite in number and coherent, is naturally equivalent to any of the following categories:

(i) $\mathrm{D^b}(\mathrm{Sym}(W^*))/\mathrm{D^b_{art}}(\mathrm{Sym}(W^*))$, which is (equivalent to) Serre's classical description.

138 (ii) $\underline{\mathrm{Mon}}^0(\mathrm{Sym}(W^*))/(\text{Koszul complexes})$, that is, the full subcategory of the category in (i) which is generated by all bounded complexes of graded free $\mathrm{Sym}(W^*)$-modules, *linear* in the sense that the module in (complex-)degree k is of the form $A_k \otimes_K \mathrm{Sym}(W^*)(k)$ with A_k a finite dimensional K-vectorspace.

REMARK. Direct sums of appropriately shifted Koszul complexes are the only such linear complexes with artinian cohomology. As morphisms between linear complexes are homotopic iff they are equal, it is clear a priori that the morphism groups in this category consists precisely of all morphisms of such complexes modulo those which factor over a direct sum of Koszul complexes. The point of the Bernšteĭn–Gelfand–Gelfand correspondence, if considered from this point of view, is hence that any complex in $\mathrm{D^b}(\mathbb{P}(W^*))$ is quasi-isomorphic to (the sheafification of) such a linear complex. For single sheaves, this is precisely the main ingredient established and used by D. Quillen in his calculation of the higher K-groups of projective bundles, [**Qui73**, §8]; see also Remark (c) in (9.5).

Dismantling such a linear complex to keep only the generating sets in each (complex-)degree—and tensoring with $\det(W)$—we get the incarnation of $\mathrm{D^b}(\mathbb{P}(W))$ as provided by the Bernšteĭn–Gelfand–Gelfand correspondence:

(iii) $\underline{\mathrm{mod}}(\bigwedge(W)) = \underline{\mathrm{MCM}}(\bigwedge(W))$, the category of all finitely generated (=maximal Cohen–Macaulay) graded modules over the exterior algebra generated by W in degree one, modulo all finitely generated projective (=free =injective) modules over this local quasi-Frobenius algebra.

(iv) $\underline{\mathrm{APC}}(\bigwedge(W))$, the homotopy category of all acyclic complexes of finitely generated, graded free $\bigwedge(W)$-modules.

(v) $\mathrm{D^b}(\bigwedge(W))/\,\mathrm{D^b_{perf}}(\bigwedge(W))$, the stable derived category of the exterior algebra over W.

The equivalence of the descriptions in (i)–(iii) and (v) is the original result in [**BGG78**]. The equivalence with (iv) has then been remarked by A. A. Beĭlinson, as S. I. Gelfand informed me—see also [**Gel84**].

CHAPTER 10

Applications to singularities and hypersurfaces

In the following R will always denote a commutative (noetherian) ring of finite 139
Krull dimension, $r = \dim R$. For simplicity (only), R will also always be assumed to have connected prime spectrum.

10.1 Local Tate cohomology. Let $(R, \mathfrak{m}, k)$ be a local commutative Gorenstein ring with maximal ideal $\mathfrak{m}$ and residue class field $k = R/\mathfrak{m}$. Then, by analogy with (8.2) and (8.5), for any integer i, we call

$$\widehat{\mathrm{H}}^i_R(M) := \underline{\mathrm{Ext}}^i_R(k, M)$$

the i-th *local Tate cohomology of* R with values in M, which may be any (complex of) R-module(s) (in $\mathrm{D}^-(\mathrm{Mod}(R))$, (7.4.1)).

As in general, $\widehat{\mathrm{H}}_R(M)$ will be a graded module over the *local Tate cohomology ring of* R, $\widehat{\mathrm{H}}_R := \widehat{\mathrm{H}}_R(k)$, with respect to the Yoneda product.

As R is local, any finitely generated R-module M admits—up to isomorphism—a unique decomposition $M = M' \oplus R^m$ where M' does not contain any free summand, (see [**Swa85**, Thm 2.6]).

If now M is already an MCM over R, the groups $\widehat{\mathrm{H}}^i_R(M)$ yield nothing new:

LEMMA 10.1.1. *Let M be an MCM over a local Gorenstein ring $(R, \mathfrak{m}, k)$, $M = M' \oplus R^m$ the decomposition in which M' contains no free summand. Then one gets*

(i) $\widehat{\mathrm{H}}^i_R(M) = \mathrm{Ext}^i_R(k, M)$ *for any* $i > r = \dim R$,
(ii) $\widehat{\mathrm{H}}^r_R(M) = \mathrm{Ext}^r_R(k, M') = \mathrm{Ext}^r_R(k, M)/k^m$,
(iii) $\widehat{\mathrm{H}}^{r-1}_R(M) = M' \otimes_R k$, *and*
(iv) $\widehat{\mathrm{H}}^i_R(M) = \mathrm{Tor}^R_{r-1-i}(M, k)$ *for any* $i < r - 1$.

PROOF. This follows immediately from (6.2.5) 3—or also directly from the construction of a complete resolution for M as in (4.5.1)—once it is observed that $\mathrm{RHom}_R(k, R) = k[-r]$ for any local Gorenstein ring and that $\mathrm{Ext}^i_R(k, M) = 0$ for any $i < r$ and all MCM over R. □

Usually, it is a rather difficult problem to determine (the dimension of) the k- 140
vectorspace occurring in this lemma. Sometimes, to know that they are necessarily modules over $\widehat{\mathrm{H}}^0_R$ will yield restrictions, for example if there are no "small" simple modules over the ring $\widehat{\mathrm{H}}^0_R$.

The size of $\widehat{\mathrm{H}}^0_R$ is essentially known from (5.5.2). More generally, we get, as in (8.5.6), the following information on the local Tate cohomology ring of R:

PROPOSITION 10.1.2. *Let $(R, \mathfrak{m}, k)$ be a local Gorenstein ring. Then*

(i) *The $\mathbb{Z}$-graded k-algebra $\widehat{\mathrm{H}}_R$ carries a natural involution, with respect to which it becomes isomorphic to its graded opposite k-algebra.*

(ii) *There exist non-degenerate k-bilinear pairings, associative for the Yoneda-product on $\widehat{\mathrm{H}}_R$:*

$$\widehat{\mathrm{H}}_R^{i+r-1} \otimes_k \widehat{\mathrm{H}}_R^{-i} \longrightarrow k \quad \text{for all } i\,,$$

in particular, $\widehat{\mathrm{H}}_R$ is a self-injective k-algebra.

(iii) *The k-algebra $\widehat{\mathrm{H}}_R^0 = \underline{\mathrm{Hom}}_R(\mathbf{M}(k), \mathbf{M}(k))$ is naturally isomorphic to its own opposite algebra. Its dimension over k is given by*
  - (a) $\widehat{\mathrm{H}}_R^0 = 0$ *if and only if R is regular,*
  - (b) $\dim \widehat{\mathrm{H}}_R^0 \geq 2^{\dim R}$ *if R is not regular, and in this case*
  - (c) $\dim \widehat{\mathrm{H}}_R^0 = 2^{\dim R}$ *if and only if either $\dim R \leq 1$ or R is an (abstract) hypersurface ring. In these cases, $\widehat{\mathrm{H}}_R^0$ is a quasi-Frobenius algebra, if $\dim R$ is odd.*

PROOF. (i), (ii) and the first assertion in (iii) follow from the general theory developed in Chapter 7. Also (iii) (a) is obvious, as $\widehat{\mathrm{H}}_R^0 = 0$ iff k is of finite projective dimension over R iff R is regular.

The given bounds on the dimension encode well-known results on the ranks of the free resolution of k over R. More precisely, one has in general:

LEMMA 10.1.3. [1] *Let $(R, \mathfrak{m}, k)$ be a local Gorenstein ring, M any MCM over R and $\boldsymbol{x} = (x_1, \ldots, x_d)$ a regular sequence in $\mathfrak{m}$. If the classes of the elements x_i are linearly independent over k in $\mathfrak{m}/\mathfrak{m}^2$, then there is an isomorphism of graded k-vectorspaces, natural in M:*

$$\underline{\mathrm{Ext}}_R(k, M) \xrightarrow{\cong} \bigwedge((\boldsymbol{x}/\boldsymbol{x}^2)^*) \otimes_{R/\boldsymbol{x}R} \underline{\mathrm{Ext}}_{R/\boldsymbol{x}R}(k, M/\boldsymbol{x}M)\,.$$

141 PROOF. The proof extends the corresponding one for the ordinary Ext's. The ring homomorphism $R \to R/\boldsymbol{x}R$ is perfect as $\boldsymbol{x}$ is a regular sequence, and the resolution of $R/\boldsymbol{x}R$ is given by a Koszul complex. Now apply the spectral sequence $''E^{i,j}$ from (7.2.3) and remark that it degenerates completely at the E_2-level iff the classes of the x_i are linearly independent modulo $\mathfrak{m}^2$. □

To conclude the proof of the proposition, one may assume—after eventually extending the residue class field—that there is a maximal regular sequence of elements whose classes in $\mathfrak{m}/\mathfrak{m}^2$ are linearly independent, so that $R/\boldsymbol{x}R$ is a zero-dimensional Gorenstein ring. Then (10.1.1) shows that $\underline{\mathrm{Ext}}_{R/\boldsymbol{x}R}^{-i}(k,k) = \mathrm{Tor}_{i-1}^{R/\boldsymbol{x}R}(k,k)$ for any $i > 0$. Hence[2]

$$\widehat{\mathrm{H}}_R^0 = \underline{\mathrm{Ext}}_R^0(k,k) = k \oplus \Big(\bigoplus_{j=1}^{r} \bigwedge^j((\boldsymbol{x}/\boldsymbol{x}^2)^*) \otimes_{R/\boldsymbol{x}R} \mathrm{Tor}_{j-1}^{R/\boldsymbol{x}R}(k,k)\Big)$$

which implies (b). It also follows that the equality in (c) holds iff all the k-vectorspaces $\mathrm{Tor}_j^{R/\boldsymbol{x}R}(k,k)$ are one-dimensional for $j < r$, which happens if and only if either r is at most 1 or the dimension of $\mathfrak{m}/\mathfrak{m}^2$ is precisely $\dim R + 1$.

---

[1]The statement of this lemma is incorrect, but it is not needed in the proof of (10.1.3); see Remark A.10.1 for a correction.

[2]This requires more detail; see Remark A.10.3.

The last assertion in (iii) is clear (from (ii)) if $\dim R \leq 1$. In case of a hypersurface, it follows from the explicit knowledge of the ordinary Yoneda-Ext-algebra of k over R.

Namely, one has explicitly

EXAMPLE 10.1.4. (i) If $(R, \mathfrak{m}, k)$ is a one-dimensional local ring which is Gorenstein, then $\widehat{\mathrm{H}}^0_R$ is one of the following k-algebras:

(a) $k \times k$

(b) a quadratic field extension $k(a)$ of k, or

(c) the local k-algebra $k[x]/x^2$.

Case (a) occurs for example for a non-singular quadric, (b) for a regular but singular quadric—see [**BEH87**]—and (c) is the generic case which occurs if the multiplicity of R is at least 3.

(ii) If $(R, \mathfrak{m}, k)$ is an (abstract) hypersurface ring, the natural map of vectorspaces $\mathrm{Sym}_2(\mathfrak{m}/\mathfrak{m}^2) \to \mathfrak{m}/\mathfrak{m}^2$ has at most a one-dimensional kernel. Let

$$Q\colon \Gamma_2((\mathfrak{m}/\mathfrak{m}^2)^*) \to k$$

be the quadratic form determined by this kernel (which may hence be zero!).

Then $\widehat{\mathrm{H}}^0_R = C^+(Q)$, the *even Clifford-algebra* of this quadratic form—and the involution on $\widehat{\mathrm{H}}^0_R$ is the main involution of the Clifford-algebra. If $Q = 0$, $C^+(Q) = \bigwedge^+((\mathfrak{m}/\mathfrak{m}^2)^*)$ of course.

If $Q \neq 0$, we get restrictions as mentioned above: The dimension of a simple module over $C^+(Q)$ is of the form 2^l, where l depends on Q—and the field k. It follows then that for any module M over R the dimension of any Tate **142**
cohomology group has to be divisible by 2^l. It follows then—for example—that an indecomposable MCM over R has to have a rank which is divisible by 2^{l-1}—see again [**BEH87**] for details. □

We conclude this section by restating the Duality Theorem 7.7.5 in the commutative context:

PROPOSITION 10.1.5. *Let $(R, \mathfrak{m}, k)$ be a local Gorenstein ring, M a maximal Cohen–Macaulay R-module which is locally free on the punctured spectrum* $\mathrm{Spec}(R) \smallsetminus \{\mathfrak{m}\}$. *If then E denotes an injective envelope of the R-module* $\mathrm{Ext}_R^{\dim R}(k, R)$, *there exists a natural R-linear trace*

$$\tau_M\colon \underline{\mathrm{Ext}}_R^{\dim R - 1}(M, M) \longrightarrow E\,,$$

such that for any complex N in $\mathrm{D^b}(R)$, *the pairings obtained from composing the Yoneda product with this trace are R-bilinear and non-degenerate:*

$$\underline{\mathrm{Ext}}_R^{\dim R-1+i}(N, M) \otimes_R \underline{\mathrm{Ext}}_R^{-i}(M, N) \xrightarrow{-\circ-} \underline{\mathrm{Ext}}_R^{\dim R-1}(M, M) \xrightarrow{\tau_M} E\,.$$

In particular, the two modules which are paired into E have the same (finite) length, and the stable Yoneda-Ext-algebra $\underline{\mathrm{Ext}}_R(M, M)$ *is self-injective.*

PROOF. Take $R = \omega_S = S = T = \omega_T$ and $M = {}^*N = N^*$ in (7.7.5) and remark that M locally free off the maximal ideal implies that it is stably transversal to any complex N. □

REMARKS. (a) If $R_\mathfrak{p}$ is regular for any prime different from $\mathfrak{m}$, any maximal Cohen–Macaulay R-module is free off the maximal ideal.

(b) Of course, $\mathrm{Ext}_R^{\dim R}(k,R) = k$ as an R-module, so that the module E is just an injective envelope of the residue class field. In particular, as explained in (7.7), if R contains a copy of k, one obtains the trace above as the R-linear extension of a k-linear trace into k.

The reason why we introduced E as an envelope of $\mathrm{Ext}_R^{\dim R}(k,R)$—and not just of k—is furnished by the graded case:

If R, instead of being local, is non-negatively graded and if the irrelevant ideal generated by all elements of positive degree is maximal, then the assertion above still holds for graded MCMs and the graded Tate cohomology. But in this case, as before in (8.5), we have to keep track of the grading of $\mathrm{Ext}_R^{\dim R}(k,R)$ to obtain a homogeneous pairing.

Let us give an explicit

143 EXAMPLE 10.1.6. Let $S = k[x_0,\dots,x_{n+d}]$ be a standard graded polynomial ring over k, $R = S/\boldsymbol{f}$ a homogeneous complete intersection of dimension $n+1$, defined by the regular sequence $f_1,\dots,f_d$ of homogeneous polynomials with $\deg(f_i) = a_i$. Then the canonical module of R over k is given by $\omega_R = R(-n-d-1+\sum a_i)$—and satisfies hence $\mathrm{Ext}_R^{n+1}(k,\omega_R) = k(0)$.

It follows that for any graded MCM R-module M which is locally free off the irrelevant ideal and for any complex N of graded R-module in $\mathrm{D^b}(R)$, there are perfect pairings

$$\underline{\mathrm{Ext}}_R^{i+n}(N, M(-n-d-1+\sum a_i)) \times \underline{\mathrm{Ext}}_R^{-i}(M,N) \longrightarrow k\,,$$

homogeneous for the total degree. In particular, taking $M = N$, the "center of symmetry" is in degree $[\frac{n}{2}](\frac{-n-d-1+\sum a_i}{2})$ for the self-dual algebra $\underline{\mathrm{Ext}}_R(M,M)$.

Finally, we come to the example which motivated our study here:

10.2 Maximal Cohen–Macaulay modules on hypersurfaces. Considering the classical case of integral group representations, the theory of hypersurfaces as compared to general (commutative) Gorenstein rings should show similar particular features as the cohomological theory of cyclic groups if compared to that of arbitrary finite groups.

After all, integral group rings of cyclic groups are "hypersurfaces" relative to the regular ring $\mathbb{Z}[t]$, and the key fact—namely that complete resolutions can be chosen to be periodic of period two—holds for general hypersurfaces by **[Eis80]**.

As an instance of this analogy, we will introduce in the next section *Herbrand differences* for any two modules over a hypersurface with only isolated singularities.

But first, we recall some essential facts and study the structure of the Tate cohomology over general hypersurfaces.

To fix the notations, P will henceforth denote a regular domain, f a non-zero element in P, and $p\colon P \to R := P/fP$ the projection onto the hypersurface ring R defined by f. We denote $\mathrm{Ker}(p) = fP$ by I and hence I/I^2 is the conormal module of p—or of R over P. It is a free R-module of rank one.

If $P = k[\boldsymbol{x}]$ is a polynomial ring, f a homogeneous element of degree d, then $I/I^2 = R(-d)$ as a graded free R-module.

144 As before $p_*\colon \mathrm{D^b}(R) \to \mathrm{D^b}(P)$ denotes the forgetful functor, p^* its left, $p^!$ its right adjoint. As P is regular, the thick hull of p^* is precisely $\mathrm{D^b_{perf}}(R)$, the category of perfect complexes over R.

The structure theorem for modules over hypersurface rings, due to D. Eisenbud, can be expressed as follows:

THEOREM 10.2.1. *With notations as just introduced, there is a distinguished triangle of endo-functors on* $\mathrm{D^b}(R)$*—or even on* $\mathrm{D}^-(\mathrm{Mod}(R))$*:*[3]

$$- \otimes_R I/I^2[1] \longrightarrow p^*p_* \xrightarrow{\pi} \mathrm{id} \xrightarrow{s} - \otimes_R I/I^2[2]\,,$$

where π *is the co-unit of the adjunction* (p^*, p_*). □

Remark that there is no need to derive the tensor product with I/I^2 as this module is already free over R. From the abstract point of view, this triangle just expresses that R is a P-module of projective dimension one, resolved by the Koszul complex $0 \to I \to P \to 0$.

Namely, if X is any complex of R-modules, p_*X is still the same complex but regarded as consisting of P-modules, and p^*p_*X is represented by the total complex associated to $X \otimes_P (0 \to I \to P \to 0)$.

Now the obvious spectral sequence associated to this double complex shows that the projection onto

$$X = (X \otimes_R R) \otimes_P R = X \otimes_P \mathrm{H}^0(0 \to I \to P \to 0)$$

—which represents $\pi(X)$—is surjective and that its kernel is quasi-isomorphic to $X \otimes_P I[1] = X \otimes_R P \otimes_P I[1] = X \otimes_R I/I^2[1]$.

But this is precisely the statement of the theorem. The difficult part, examined in [**Eis80**], is to give s explicitly for a complex X.

Here we will only show how this theorem implies immediately that resolutions over R become eventually periodic of period two:

If X is any bounded above complex of projective R-modules with finitely generated and bounded cohomology, p^*p_*X is perfect. Hence there exists a finite free complex representing it. But this means that s becomes a quasi-isomorphism after truncating those finitely many degrees in which this representative of p^*p_*X is concentrated.

If P is local, X a minimal complex of finitely generated free R-modules then **145**
X is determined up to isomorphisms of complexes in its isomorphism class in $\mathrm{D}^-(\mathrm{Mod}(R))$—and similarly in case that P is non-negatively graded with a local regular ring in degree zero.

In these cases, $X \otimes_R I/I^2$ will also be minimal and hence $s(X)$ an isomorphism of complexes—except at a finite number of terms. This proves that X—up to the twist with the free module I/I^2—can be chosen to be eventually periodic of period two. Remark that if X is the resolution of a single R-module M, it can be chosen to be periodic after at most $\mathrm{projdim}_P\, p_*M$ many steps.

REMARK. (a) The complex $I/I^2[1]$ is the natural representative of the relative cotangent complex $\mathbb{L}_{R/P}$ of R over P, so that s can be thought of as a natural transformation from the identity on $\mathrm{D}^-(\mathrm{Mod}(R))$ into the functor $- \otimes_R \mathbb{L}_{R/P}[1]$. But, following Illusie's treatment of Atiyah-classes, [**Ill71**, IV.2.3], the relative Atiyah-class, $\mathrm{at}_{R/P}$, also provides such a transformation, and indeed one sees easily from the explicit construction of s in [**Eis80**] that they are the same.

[3]The functor s is Eisenbud–Gulliksen operator usually denoted χ.

Now we will show that (10.2.1) determines completely the natural transformation $c\colon \mathrm{Ext}_R(-,-) \to \underline{\mathrm{Ext}}_R(-,-)$ from the ordinary to the stabilized Ext-groups, (6.2.5) 3.

As p^*p_* maps $\mathrm{D^b}(R)$ into $\mathrm{D^b_{perf}}(R)$, the class of the transformation s becomes an isomorphism of functors in the stable derived category $\underline{\mathrm{D}}^{\mathrm{b}}(R)$.

If e denotes a (homogeneous) generator of $(I/I^2)^*$, the normal module of R over P—which is still free of rank one—we may "normalise" s to the morphism of functors $s' := (1 \otimes_R e) \circ s\colon \mathrm{id} \to \mathrm{T}^2$.

Then (10.2.1) yields homomorphisms of graded rings—$\deg s' = 2$:

$$\mathbb{Z}[s'] \longrightarrow \bigoplus \mathrm{Hom}_{\mathrm{D^b}(R)}(\mathrm{id}, \mathrm{T}^i) \longrightarrow \bigoplus \mathrm{Hom}_{\underline{\mathrm{D}}^{\mathrm{b}}(R)}(\mathrm{id}, \mathrm{T}^i)\,, \tag{10.2.2}$$

which, when evaluated in a complex X, become the ring homomorphisms

$$\mathbb{Z}[s'] \longrightarrow \mathrm{Ext}_R(X,X) \xrightarrow{c(X,X)} \underline{\mathrm{Ext}}_R(X,X)\,,$$

sending s' into the class of $s'(X)$.

To say that s' is a natural transformation of functors, amounts to the fact that the image of s' is central[4] and that the induced right or left actions on $\mathrm{Ext}_R(X,Y)$ for any two complexes X, Y in $\mathrm{D^b}(R)$ by the Yoneda product with $s'(X)$ and $s'(Y)$ resp. are the same, so that any of the graded Ext-groups becomes a symmetric $\mathbb{Z}[s']$-bimodule.

As $c \circ s'$ is invertible and c is an isomorphism in large degrees:

146 THEOREM 10.2.3. *For any two (complexes of) R-modules M and N (in $\mathrm{D^b}(R)$), one has*

$$\underline{\mathrm{Ext}}_R(M,N) = \mathrm{Ext}_R(M,N) \otimes_{\mathbb{Z}[s']} \mathbb{Z}[s', s'^{-1}]\,,$$

and the natural transformation $c(M,N)$ is the inclusion of the first factor. Its kernel is hence precisely the s'-torsion in $\mathrm{Ext}_R(M,N)$. □

REMARK. (b) As s' is of degree 2, $\underline{\mathrm{Ext}}_R(M,N)$ is in fact already completely determined by the homogeneous localizations

$$\underline{\mathrm{Ext}}^{+}_R(M,N) = \Big(\bigoplus_i \mathrm{Ext}^{2i}_R(M,N)\Big)/(s'-1)\,,$$

which equals any $\underline{\mathrm{Ext}}^{2j}_R(M,N)$, and

$$\underline{\mathrm{Ext}}^{-}_R(M,N) = \Big(\bigoplus_i \mathrm{Ext}^{2i+1}_R(M,N)\Big)/(s'-1)\,,$$

which equals any $\underline{\mathrm{Ext}}^{2j+1}_R(M,N)$, whence we will call these groups also the "even" and "odd" Tate cohomology respectively.

In other words, $\underline{\mathrm{D}}^{\mathrm{b}}(R)$ is in fact a $\mathbb{Z}/2\mathbb{Z}$-graded category.

Now assume (for simplicity) that P contains a field k and that f has only isolated singularities (non-regular points will suffice[5]).

Then $\mathrm{Ext}_R(M,N)$ will be a finitely generated $\mathbb{Z}/2\mathbb{Z}$-graded module over $k[s']$—as $\deg s' = 2$—and can hence be decomposed into the even and odd submodules, E^+ and E^- respectively. Furthermore, the s'-torsion of either $E^{+/-}$ can be characterized in the usual way, say by means of the Jordan-blocks of the action of s.

---

[4]This amounts to the statement that the image of s' is in the center of $\mathrm{D^b}(R)$

[5]That is to say the non-closed points of P/f are regular rather than smooth.

Remain the free $k[s']$-modules $E^{\pm}$ module torsion, which are classified by their ranks. The difference of these ranks behaves like a "stable" Euler–Poincaré-characteriztic and seems hence to be a very reasonable invariant—in particular if compared to the case of cyclic groups—whence we give it a name:

10.3 The Herbrand difference.

DEFINITION 10.3.1. Let $(R, \mathfrak{m}, k)$ be a local hypersurface ring with $\mathfrak{m}$ the only prime at which R is not regular.

For any two (complexes of) R-modules M and N set:

$$\mathrm{h}(M,N) := \mathrm{length}(\underline{\mathrm{Ext}}_R^+(M,N)) - \mathrm{length}(\underline{\mathrm{Ext}}_R^-(M,N))$$

and call this number the *Herbrand difference* of (M, N) over R.[6] **147**

REMARKS. (a) If R is not local, but f still has only isolated non-regular points, one may define these differences at any non-regular point (and sum up) to get *local* (and *global*) Herbrand differences.

This applies in particular for homogeneous hypersurfaces, which are only non-regular at the origin. In this case, it is to be understood that we sum up over all the graded components of the corresponding Ext's.

(b) The name is apparently motivated by the *Herbrand quotient* for cyclic groups, [**Ser79**, VIII, §5], defined as

$$\mathrm{h}(M) := |\,\widehat{\mathrm{H}}^0(G,M)|/|\,\widehat{\mathrm{H}}^1(G,M)|\,,$$

the quotient of the orders of the Tate cohomology groups for a given representation M of the cyclic group G. Comparing with our definition,

$$\mathrm{h}(M) = \prod_p p^{\mathrm{h}_p(\mathbb{Z},M)}$$

where h_p is the local Herbrand difference at the prime p. (Of course only those primes are non-regular—and hence contribute—which divide the order of G.)

(10.3.2) Due to the periodicity of Tate cohomology, for any R-module M and any short exact sequence $0 \to N_1 \to N_2 \to N_3 \to 0$ there is an exact hexagon

$$\begin{array}{ccccc}
\underline{\mathrm{Ext}}_R^+(M,N_1) & \longrightarrow & \underline{\mathrm{Ext}}_R^+(M,N_2) & \longrightarrow & \underline{\mathrm{Ext}}_R^+(M,N_3) \\
\uparrow & & & & \downarrow \\
\underline{\mathrm{Ext}}_R^-(M,N_1) & \longleftarrow & \underline{\mathrm{Ext}}_R^-(M,N_2) & \longleftarrow & \underline{\mathrm{Ext}}_R^-(M,N_3)
\end{array}$$

and similarly for exact sequences in the first argument. It follows:

(i) $\mathrm{h}(-,-)$ is additive on short exact sequences in each argument, and also on distinguished triangles in $\mathrm{D^b}(R)$.

This implies that h induces a $\mathbb{Z}$-valued, bilinear pairing on the Grothendieck group $\mathrm{G}_0(R)$ of all finitely generated R-modules. This pairing will be denoted by the same symbol.

[6] Up to a sign this is the θ-invariant; see [**68**] and also [**38**].

(ii) $\mathrm{h}(U,-) = \mathrm{h}(-,U) = 0$ for all finitely generated R-modules of finite projective dimension (or U in $\mathrm{D}^{\mathrm{b}}_{\mathrm{perf}}(R)$).

148 This is clear from the definition of Tate cohomology and shows that h as a pairing on $\mathsf{G}_0(R)$ contains $\mathsf{K}_0(R)$, the Grothendieck group of all finitely generated projective R-modules, in its radical. Hence it induces a pairing on $\underline{\mathsf{K}}_0(R) := \mathsf{G}_0(R)/\mathsf{K}_0(R)$, (4.9), still to be denoted h.

Next we want to apply the Duality Theorem in connection with the periodicity. Let us first state this as a separate

COROLLARY 10.3.3. *Let $(R, \mathfrak{m}, k)$ be a local hypersurface ring with $\mathfrak{m}$ as its only non-regular prime. Then*

(i) *If $\dim R$ is even, $\underline{\mathrm{Ext}}^+_R(M,N)$ and $\underline{\mathrm{Ext}}^-_R(N,M)$ are dual to each other. In particular, they have the same length.*
(ii) *If $\dim R$ is odd, $\underline{\mathrm{Ext}}^+_R(M,N)$ is dual to $\underline{\mathrm{Ext}}^+_R(N,M)$ and the same holds for the odd Tate cohomology groups.*
In particular,
(iii) *For any complex M in $\mathrm{D}^{\mathrm{b}}(R)$, the $\mathbb{Z}/2\mathbb{Z}$-graded stable Ext-algebra*

$$\underline{\mathrm{Ext}}^+_R(M,M) \oplus \underline{\mathrm{Ext}}^-_R(M,M)$$

is quasi-Frobenius and carries a graded involution. If $\dim R$ is odd, already $\underline{\mathrm{Ext}}^+_R(M,M)$ is a quasi-Frobenius subalgebra.

PROOF. This is obvious from (10.1.5). □

In the graded case, we can still do a little bit better:

COROLLARY 10.3.4. *Let R be homogeneous quotient of a polynomial ring in an even number of variables by a polynomial of odd degree d—and with the origin as only non-regular point.*

Then for any graded (complex of) R-module(s) M (in $\mathrm{D}^{\mathrm{b}}(R)$), the length of $\underline{\mathrm{Ext}}^{\pm}_R(M,M)$ is even.

PROOF. Just substitute the given data in (10.1.6) to obtain that the symmetry is centerd at half an integer. □

Applying all this to the Herbrand differences, we get

(iii) $\mathrm{h}(M,N) = (-1)^{\dim R-1}\,\mathrm{h}(N,M)$ for all (complex of) R-modules M and N (in $\mathrm{D}^{\mathrm{b}}(R)$).

In particular, the pairing given by h is

- *alternating* if $\dim R$ is even,
- *symmetric* if $\dim R$ is odd, and
- 149 the quadratic form $Q_{\mathrm{h}}(X) := \mathrm{h}(X,X)$ is *even*, if R is defined by a polynomial of odd degree.

Before we go on to deduce further properties of h, let us make the

REMARK. (a) The specialisation of the Duality Theorem to hypersurfaces shows that for any complex (or just MCM) on a hypersurface the stable $\mathbb{Z}/2\mathbb{Z}$-graded endomorphism ring shares the characteriztic properties of the Clifford-algebra of a quadratic form—and it is a Clifford-algebra in case of the (MCM-approximation of the) residue class field as mentioned in (10.1.4) (ii). A satisfactory explanation is missing—except when the hypersurface itself is a regular quadric.

Coming back to Herbrand differences, let us translate the existence of the "trivial" duality, (4.6):

(iv) Let M, N be two maximal Cohen–Macaulay R-modules (or complexes in $\mathrm{D^b}(R)$) and set $(-)^* = \mathrm{Hom}_R(-, R)$, (resp. $= \mathrm{RHom}(-, R)$). Then one has obviously

$$\mathrm{h}(M, N) = \mathrm{h}(N^*, M^*)\,.$$

(v) Let L be a perfect complex of R-modules, l its Euler–Poincaré characteriztic. Then it follows by induction on the number of non-zero components of L that

$$\mathrm{h}(- \otimes^{\mathsf{L}}_R L, -) = l \cdot \mathrm{h}(-, -)\,.$$

(vi) If A is (a syzygy module of) some artinian R-module, then

$$\mathrm{h}(A, -) = \mathrm{h}(-, A) = 0\,.$$

This follows for artinian modules themselves from the fact that their classes in $\mathsf{G}_0(R)$ are torsion—as for any local Cohen–Macaulay ring—but $\mathrm{h}(A, -)$ is an additive function into the torsion-free group of integers. For syzygy modules, the statement follows from (ii).

Finally let us mention a different description of h in case that R is normal:

(vii) Let N and M be two MCMs over a normal hypersurface ring. Then

$$\mathrm{h}(M, N) = \mathrm{h}^1_{\mathfrak{m}}(N \otimes_R M^*) - \mathrm{h}^0_{\mathfrak{m}}(N \otimes_R M^*)$$

where $\mathrm{h}^i_{\mathfrak{m}}(-)$ denotes the length of the local cohomology module $\mathrm{H}^i_{\mathfrak{m}}(-)$.

This is obtained from the exact sequence in (6.4.1) (iii), as under the assump- **150**
tions made $\mathrm{Hom}_R(M, N)$ is reflexive, whence

$$\mathrm{H}^i_{\mathfrak{m}}(\mathrm{Hom}_R(M, N)) = 0 \quad \text{for } i = 0, 1$$

and hence

$$\mathrm{H}^i_{\mathfrak{m}}(N \otimes_R M^*) = \underline{\mathrm{Ext}}^{i-1}_R(M, N) \quad \text{for } i = 0, 1.$$

To summarise the properties of the Herbrand difference, let us introduce—aside $\mathsf{K}_0(R)$ and $\mathsf{G}_0(R)$—also the Grothendieck group $\mathsf{K}_0(\mathrm{art}(R))$ of all artinian R-modules. Then summing up over all the non-regular points we get:

THEOREM 10.3.5. *Let R be a hypersurface ring which has only isolated non-singular*[7] *points and set $\varepsilon := (-1)^{\dim R - 1}$.*

Then the Herbrand difference h *induces an ε-symmetric pairing on*

$$\mathbb{K}(R) := \mathsf{G}_0(R)/(\mathsf{K}_0(R) + \mathsf{K}_0(\mathrm{art}(R)))\,. \qquad \square$$

This pairing is ε-invariant under the duality induced by $\mathrm{RHom}_R(-, R)$ and $\mathsf{K}_0(R)$-bilinear with respect to the natural $\mathsf{K}_0(R)$-module structure on $\mathsf{G}_0(R)$.

Formulating it separately in the graded case, we get

THEOREM 10.3.6. *Let R be a homogeneous hypersurface ring defined by a polynomial of degree d in $k[x_0, \ldots, x_n]$. Let X be the underlying projective hypersurface in $\mathbb{P}^n$ and assume that it is regular.*

Then the Herbrand difference induces a bi-linear pairing on

$$\mathbb{K}(X) := \mathsf{K}_0(X)/\mathsf{K}_0(\mathbb{P}^n)\,,$$

[7] As usual, non-regular suffices.

which is symmetric if $\dim X = n - 1$ *is even; and the associated quadratic form is even if furthermore the degree* d *is odd. In case that* $\dim X$ *is odd, the pairing is alternating.* □

This theorem suggests that the Herbrand difference should serve as the (?) intersection form on the (co-)primitive cycles of the projective hypersurface X.

Let us consider some examples:

151 EXAMPLE 10.3.7. Let X be a non-singular quadric. Then it is known, by [**Knö87**], see also [**BEH87**], that there is precisely one non-free and indecomposable graded MCM—up to shifts and isomorphisms—if $\dim X$ is odd. In this case, $\underline{\mathsf{K}}_0(R) = \mathbb{Z}/2\mathbb{Z}$ and hence h is trivial.

If $\dim X$ is even, there are precisely two such MCMs and the group $\underline{\mathsf{K}}_0(R)$ is free of rank one, generated by either one of them. For any of these generating modules M, one has $\mathrm{h}(M, M) = 1$.

EXAMPLE 10.3.8. Let R be a simple analytic hypersurface singularity which is even-dimensional. Then $\underline{\mathsf{K}}_0(R)$ is finite and hence h is trivial again.

In case of odd dimension—and characteriztic different from 2—one is reduced to the curve case by H. Knörrer's periodicity result, [**Knö87**]. But for a reduced plane curve singularity, the Grothendieck group $\mathsf{G}_0(R)$ is generated—at least up to torsion—by the structure sheaves of the irreducible branches, and it is left as an exercise to calculate the pairing induced by h on these modules.

Finally we consider:

10.4 The 27 lines on a cubic—once again. Let $P = k[x, y, u, v]$ and let f be a homogeneous polynomial of degree three such that the underlying surface X is smooth.

(10.4.1) A *line* on X is apparently given by two linear forms l_1, l_2 such that f can be written as

$$f = l_1 q_1 + l_2 q_2$$

for two quadratic polynomials q_1, q_2 in P.

Set $A = P/(l_1, l_2)$ and let $L = \mathrm{Proj}(A)$, and denote $A' = P/(l_1', l_2')$ and $L' = \mathrm{Proj}(A')$ a (module of a) second line.

(10.4.2) Let $(-,-)$ denote the intersection form on $\mathrm{Pic}(X)$. Then it is well-known that $\mathrm{Pic}(X)$ looks as follows:

$$\mathrm{Pic}(X) = \bigoplus_{i=0}^{6} \mathbb{Z} \cdot l_i$$

with

$$(l_i, l_j) = \begin{cases} 1 & \text{for } i = j = 0 \\ -1 & \text{for } i = j > 0 \\ 0 & \text{for } i \neq j. \end{cases}$$

Recall also that in this representation the class of the dualizing sheaf is $\omega =$
152 $(-3, 1, \ldots, 1)$. It generates the image of $\mathrm{Pic}(\mathbb{P}^3)$ in $\mathrm{Pic}(X)$. As X is two-dimensional,

$\mathsf{K}_0(X) = \mathbb{Z} \oplus \mathrm{Pic}(X)$ and hence the group on which the Herbrand difference will be defined is

$$\mathbb{K}(X) = \mathsf{K}_0(X)/\mathsf{K}_0(\mathbb{P}^3) = \mathrm{Pic}(X)/\operatorname{Im}(\mathrm{Pic}(\mathbb{P}^3) \to \mathrm{Pic}(X)) = \mathrm{Pic}(X)/\mathbb{Z}\cdot\omega\,.$$

As the classes $l_1, \ldots, l_6$ correspond to lines, the classes of lines will generate $\mathbb{K}(X)$ as an abelian group and it will be enough to determine the pairing h on pairs of the form (A, A'), and to compare it with the values $(\underline{L}, \underline{L}')$, where $\underline{L}$ and $\underline{L}'$ are the class of the corresponding line.

(10.4.3) Now we have the following table:[8]

position of lines	skew	transversal	identical
$(\underline{L}, \underline{L}')$	0	1	-1
$\underline{\mathrm{Ext}}^+_R(A, A')$	k	0	$P/(l_1q_1 + l_2q_2)$
$\underline{\mathrm{Ext}}^-_R(A, A')$	0	length 2	0
$\mathrm{h}(A, A')$	1	-2	4

This table can be verified by means of direct calculations. It follows

PROPOSITION 10.4.5. *The Herbrand difference of the homogeneous coordinate ring of a smooth cubic satisfies:*

$$\mathrm{h}(A, A') = -\frac{1}{3}(3\underline{L} + \omega, 3\underline{L}' + \omega)\,,$$

whence $(\mathsf{K}_0(X)/\mathsf{K}_0(\mathbb{P}^3) \otimes_{\mathbb{Z}} \mathbb{R}, \mathrm{h} \otimes 1)$ *is isometric as an Euclidean space to the orthogonal complement of* ω *in* $(\mathrm{Pic}(X) \otimes_{\mathbb{Z}} \mathbb{R}, -(\ ,\ ))$. □

Of course, there should exist a more conceptual proof for this.

[8] Here $R = P/f$.

APPENDIX A

Comments and errata

We, the editors, have chosen to leave the text of Buchweitz's manuscript unaltered except for its typesetting, leaving in the mysterious remarks and the few inaccuracies it contained. Footnotes have been added to the text to make clarifications and correct minor errors. In this appendix we record the more substantial corrections.

Chapter 1.

REMARK A.1.1. This comment concerns Remark 1.3 (a). According to Corollary 6.3.4, when S is strongly Gorenstein there are natural isomorphisms

$$\operatorname{Hom}_{\underline{\mathrm{D}}^{\mathrm{b}}(S)}(X, \mathrm{T}^i Y) \cong \operatorname{Hom}_{\mathrm{D}^{\mathrm{b}}(S)}(X, \mathrm{T}^i Y) \tag{A.1.2}$$

for $i \gg 0$—depending on X and Y. However, this fails in general for rings which are not strongly Gorenstein, and the isomorphisms are not a consequence of Lemma 1.2.1 as suggested in the text.

For commutative rings there is a strong converse. Assume that R is a commutative noetherian local ring with residue field k. By a result of Avramov and Veliche [**18**, Corollary 6.3], if we set $X = k$ and Y to be any module of infinite projective dimension, and if the isomorphism (A.1.2) holds for even a single i, then R must be Gorenstein; see (C.5.9) for further remarks.

Chapter 4.

REMARK A.4.1. This comment concerns the proof of Theorem 4.4.1. The argument given for part (v) is incomplete because Verdier's criterion, recalled in Theorem B.4.1, does not apply as stated. In Lemma B.4.2 we prove a generalisation of Verdier's criterion; with this in hand the proof can be completed along lines of the main text.

COROLLARY A.4.2 (Part (v) of the proof of Theorem 4.4.1). *The functor*

$$\underline{\sigma}_{\leqslant} \colon \underline{\mathrm{APC}}(S) \to \underline{\mathrm{D}}^{\mathrm{b}}(S)$$

is fully-faithful.

PROOF. Let A and B be in $\underline{\mathrm{APC}}(S)$. There is an isomorphism

$$\operatorname{Hom}_{\mathrm{K}(S)}(A, B) \cong \lim_i \operatorname{Hom}_{\mathrm{K}(S)}(A, \sigma_{\leqslant i} B)$$

by Lemma B.2.1, and there is a further isomorphism

$$\lim_i \operatorname{Hom}_{\mathrm{K}(S)}(A, \sigma_{\leqslant i} B) \cong \lim_i \operatorname{Hom}_{\underline{\mathrm{D}}^{\mathrm{b}}(S)}(\underline{\sigma}_{\leqslant} A, \underline{\sigma}_{\leqslant i} B)$$

by Lemma B.4.3—the proof of which used the generalisation of Verdier's criterion. By definition of the stable truncation functors there is an isomorphism

$$\lim_i \operatorname{Hom}_{\underline{\mathrm{D}}^{\mathrm{b}}(S)}(\underline{\sigma}_{\leqslant} A, \underline{\sigma}_{\leqslant i} B) \cong \operatorname{Hom}_{\underline{\mathrm{D}}^{\mathrm{b}}(S)}(\underline{\sigma}_{\leqslant} A, \underline{\sigma}_{\leqslant} B)\,.$$

The composition of these three isomorphisms coincides with the application of $\underline{\sigma}_{\leqslant} : \underline{\mathrm{APC}}(S) \to \underline{\mathrm{D}}^{\mathrm{b}}(S)$, so we are done. □

Chapter 5.

REMARK A.5.1. In this remark we add some detail to the proof of Lemma 5.1.1. For the "only if" part of (iv) consider the exact sequence

$$0 \to \Omega M \to P \to M \to 0$$

with P projective. In the case $\operatorname{projdim}_S M \leq 1$, applying $\operatorname{Hom}_S(M, -)$ gives the exact sequence

$$0 \to \operatorname{Hom}_S(M, \Omega M) \to \operatorname{Hom}_S(M, P) \to \operatorname{Hom}_S(M, M) \to 0\,.$$

Thus there exists $M \to P$ for which $M \to P \to M$ is the identity map. This makes M a summand of P, hence projective.

By induction on $\operatorname{projdim}_S M$, this argument allows us to conclude that M is projective as soon as $\operatorname{projdim}_S M < \infty$.

REMARK A.5.2. In the proof of Theorem 5.1.2 the following fact is used without comment: If S is strongly Gorenstein and $\varphi\colon X \to Y$ is a quasi-isomorphism of complexes of projective S-modules, then as well $\varphi^*\colon Y^* \to X^*$ is a quasi-isomorphism. (In the notation of Theorem 5.1.2, this is applied to the resolution $\varphi\colon G \to F^*$.)

By taking the cone of φ, this fact is equivalent to: If S is strongly Gorenstein and A is an acyclic complex of projective S-modules, then as well A^* is acyclic.

In this form the fact is stated in 4.6. It can be proven in a similar way to Theorem 4.4.1 part (1). Indeed, to show that $\mathrm{H}^i(A^*) = 0$, one takes j such that $i - j > \operatorname{vdim}(S)$, then

$$\mathrm{H}^i(A^*) = \operatorname{Ext}_S^{i-j}(\Omega_{-j}A, S) = 0\,.$$

Chapter 6.

REMARK A.6.1. In this remark we add some detail to the proof of Lemma 6.1.2 (ii). As explained there

$$\operatorname{Hom}_{\underline{\mathrm{D}}^{\mathrm{b}}(S)}(M, \mathrm{T}^i N) \cong \operatorname{Hom}_{\underline{\mathrm{D}}^{\mathrm{b}}(S)}(\underline{\sigma}_{\leqslant}\mathbf{CR}(M), \mathrm{T}^i N)\,,$$

and by Lemma B.4.3 there is an isomorphism

$$\operatorname{Hom}_{\underline{\mathrm{D}}^{\mathrm{b}}(S)}(\underline{\sigma}_{\leqslant}\mathbf{CR}(M), \mathrm{T}^i N) \cong \operatorname{Hom}_{\mathrm{K}(S)}(\mathbf{CR}(M), \mathrm{T}^i N)\,,$$

and then it becomes clear that this can be computed by truncating $\mathbf{CR}(M)$ at any large enough degree.

REMARK A.6.2. This remark concerns Lemma 6.2.1 (i) (b), which refers to part (v) in the proof of Theorem 4.4.1. Since the latter argument was incomplete, some clarification is needed here.

Since X is an acyclic complex of projectives and the injective dimension of S as a right module over itself is finite, X is *totally acyclic*, that is, X^* is acyclic as well; see Remark A.5.2. Therefore by Lemma B.2.1 the natural map

$$\operatorname{Hom}_S(X, Y) \to \operatorname{Hom}_S(X, \sigma_{\leqslant i} Y)$$

is a quasi-isomorphism for all i. But Y is bounded below, so $\sigma_{\leqslant i} Y = 0$ for large enough i. This makes it clear that $\operatorname{Hom}_S(X, Y) \simeq 0$.

Chapter 7.

REMARK A.7.1. This remark is about part (iii) of Proposition 7.6.3. To prove this an additional hypotheses is needed in the definition of duality morphism of finite type in Definition 7.6.2:

(iv) S is supported at some point $\mathfrak{m}$ of $\operatorname{Spec}(T)$ for which $\operatorname{height}(\mathfrak{m}) = \dim T$.

This mild condition holds automatically when T is equidimensional, or when S is supported at all of the maximal ideals of T, for instance.

In the proof of Proposition 7.6.3 (iii) it is shown that

$$\operatorname{Ext}^i_T(f_*(-), T) \cong \operatorname{Ext}^{i-d}_S(-, \omega_{S/T}).$$

Assuming condition (iv) above, one can choose a maximal ideal $\mathfrak{m}$ of T with height $\dim T$ such that $f_*(S \otimes_T T/\mathfrak{m})$ is a *non-zero* finite direct sum of copies of $T/\mathfrak{m}$. Since T is Gorenstein $\operatorname{Ext}^i_T(T/\mathfrak{m}T, T) \neq 0$ if and only if $i = \dim T$. Therefore, substituting $S \otimes_T T/\mathfrak{m}$ into the above isomorphism yields

$$\operatorname{injdim}_S \omega_{S/T} \geq \operatorname{injdim}_T T - d = \dim T - d.$$

Without the additional hypothesis (iv) above, Proposition 7.6.3 (iii) is false in general. For example, if we take $S = k$ to be a field, $T = k \times k[\![x]\!]$, and $f\colon T \to S$ the natural projection onto the first factor, then the conditions in Definition 7.6.2 are all satisfied, while $d = \operatorname{projdim}_T f_*S = 0 \neq \dim T - \operatorname{vdim} S = 1$.

REMARK A.7.2. Theorem 7.8.3 holds under the weaker hypothesis that T is regular—that is, $T_\mathfrak{p}$ is a regular local ring for each prime $\mathfrak{p}$ of T—it is not necessary to assume that the Krull dimension of T is finite. Over such a ring, every object of $\mathrm{D^b}(T)$ is perfect.

The key point is that, for a given bounded complex of finitely generated modules, being perfect is a *local property*. This is due to Bass [**Bas68**, III.6.6] for modules; and see [**14**, 4.1] for the generalisation to complexes.

Chapter 9.

REMARK A.9.1. This remark concerns the following claim made in ((9.3.2)): *The full subcategory* $\mathrm{D^b_{art}}(\mathbf{U})$ *of* $\mathrm{D^b}(\mathbf{U})$*, of those complexes with artinian cohomology, is equivalent to* $\mathrm{D^b}(\operatorname{art}(\mathbf{U}))$.

This statement follows from the work in Verdier's thesis. Indeed, taking $M = \operatorname{art}(\mathbf{U})$ and $\mathcal{A} = \operatorname{mod}(\mathbf{U})$ in [**124**, III.2.4.1 (b)] it follows the natural functor

$$\mathrm{D^b}(\operatorname{art}(\mathbf{U})) \longrightarrow \mathrm{D^b}(\mathbf{U})$$

is fully faithful and has $\mathrm{D^b_{art}}(\mathbf{U})$ as its essential image. We note that the hypotheses (E_1), (E_2) and (E_3) of [**124**, III.2.4.1 (b)] are not satisfied in this context, but in fact the *opposite* conditions are.

The key point is that $(\mathrm{E}_2^{\mathrm{op}})$ holds: For every $\mathbf{U}$-linear monomorphism $L \to X$ from an object L of $\operatorname{art}(\mathbf{U})$ to an object X of $\operatorname{mod}(\mathbf{U})$ there is a map $X \to L'$, with L' an object of $\operatorname{art}(\mathbf{U})$, such that the composition $L \to L'$ is a monomorphism; compare with [**124**, III.2.2.1].

REMARK A.9.2. In this remark we discuss the formula (9.4.10), which reads

$$\mathbf{d}(M \otimes_K N) = \mathbf{d}(M) \otimes_{\mathbf{U}(\mathbf{D}(V,W))} \mathbf{d}(N).$$

Recall that $\mathbf{U}(V,W) = \operatorname{Sym}(V) \otimes_K \bigwedge(W)$, graded naturally by $\mathbb{Z} \times \mathbb{Z}$, and

$$\mathbf{U}(\mathbf{D}(V,W)) = \mathbf{U}(W^*, V^*)$$

is the Koszul dual K-algebra. In the formula above M and N are graded $\mathbf{U}(V,W)$-modules such that their tensor product over K is locally finite for the total degree.

Theorem 9.4.2 establishes an equivalence of triangulated categories

$$\mathbf{d}\colon \mathrm{D^b}(\mathbf{U}(V,W)) \xrightarrow{\cong} \mathrm{D^b}(\mathbf{U}(\mathbf{D}(V,W)))\,.$$

The formula (9.4.10) asserts that this is a monoidal functor, or in other words, that it is an equivalence of tensor-triangulated categories. On the domain $\mathrm{D^b}(\mathbf{U}(V,W))$ one considers the tensor product over K, using the natural Hopf algebra structure of the universal enveloping algebra $\mathbf{U}(V,W)$; on the codomain $\mathrm{D^b}(\mathbf{U}(\mathbf{D}(V,W)))$ one considers the tensor product over $\mathbf{U}(\mathbf{D}(V,W))$.

In the special case that $V = 0$ one recovers the classical Bernšteĭn–Gelfand–Gelfand correspondence

$$\underline{\mathrm{MCM}}(\bigwedge(W)) \xrightarrow{\cong} \mathrm{D^b}(\mathrm{Proj}\text{-}\mathrm{Sym}(W^*))\,;$$

c.f. Proposition 9.5.1. The fact that this equivalence is compatible with the monoidal structure is interesting; even today it is difficult to find a reference treating this.

Chapter 10.

REMARK A.10.1. This remark concerns Lemma 10.1.3, which is incorrect as stated. It can be corrected as follows:

LEMMA A.10.2. *Let $(R,\mathfrak{m},k)$ be a local Gorenstein ring, M any MCM over R and $\boldsymbol{x} = (x_1,\dots,x_d)$ a regular sequence in $\mathfrak{m}$. There is an isomorphism of graded k-vectorspaces, natural in M:*

$$\underline{\mathrm{Ext}}^i_R(k,M) \cong \underline{\mathrm{Ext}}^{i+d}_{R/\boldsymbol{x}R}(k,M/\boldsymbol{x}M)\,.$$

PROOF. The formula follows from the stabilized tensor-hom adjunction associated to the $R/\boldsymbol{x}R$-R-bimodule $R/\boldsymbol{x}R$, presented in (7.2.1) and (7.2.2). As required by the discussion there, $R/\boldsymbol{x}R$ is perfect on both sides, as is its dual $\mathrm{RHom}_R(R/\boldsymbol{x}R,R)$. This yields the natural isomorphism

$$\underline{\mathrm{Ext}}_R(k,M) \cong \underline{\mathrm{Ext}}_{R/\boldsymbol{x}R}(k,\mathrm{RHom}_R(R/\boldsymbol{x}R,M))\,.$$

Since M is MCM the sequence $\boldsymbol{x}$ is M-regular, and

$$\mathrm{RHom}_R(R/\boldsymbol{x}R,M) \cong M/\boldsymbol{x}M[-d]$$

using the Koszul resolution of $R/\boldsymbol{x}R$ over R. Combining these isomorphisms yields the claim. □

REMARK A.10.3. This remark concerns the proof of Proposition 10.1.2. Since Lemma 10.1.3 cannot be applied, one must justify the displayed formula, reproduced here:

$$\text{(A.10.4)}\qquad \widehat{\mathrm{H}}^0_R = \underline{\mathrm{Ext}}^0_R(k,k) = k \oplus \Big(\bigoplus_{j=1}^{r} \bigwedge\nolimits^{j}((\boldsymbol{x}/\boldsymbol{x}^2)^*) \otimes_{R/\boldsymbol{x}R} \mathrm{Tor}^{R/\boldsymbol{x}R}_{j-1}(k,k)\Big)\,.$$

Here $\boldsymbol{x}$ is a maximal regular sequence in the maximal ideal $\mathfrak{m}$ of R whose classes in $\mathfrak{m}/\mathfrak{m}^2$ are linearly independent, so that $R/\boldsymbol{x}R$ is a zero-dimensional Gorenstein ring.

As in the proof of Lemma A.10.2 we have

$$\underline{\mathrm{Ext}}_R(k,k) \cong \underline{\mathrm{Ext}}_{R/\boldsymbol{x}R}(k,\mathrm{RHom}_R(R/\boldsymbol{x}R,k))\,,$$

cf. (7.2.1) & 7.2.2. Using the Koszul resolution of $R/\boldsymbol{x}R$ over R one obtains a quasi-isomorphism of $R/\boldsymbol{x}R$-complexes

$$\mathrm{RHom}_R(R/\boldsymbol{x}R, k) \simeq \bigoplus_{j=0}^{r} \bigwedge^{j} (\boldsymbol{x}/\boldsymbol{x}^2)^*[-j]\,.$$

Combining the last two isomorphisms yields

$$\widehat{\mathrm{H}}_R^0 = \underline{\mathrm{Ext}}_R^0(k,k) = k \oplus (\bigoplus_{j=1}^{r} \bigwedge^{j} ((\boldsymbol{x}/\boldsymbol{x}^2)^*) \otimes_{R/\boldsymbol{x}R} \mathrm{Ext}_{R/\boldsymbol{x}R}^{-j}(k,k))\,.$$

Finally, Lemma 10.1.1 shows that $\underline{\mathrm{Ext}}_{R/\boldsymbol{x}R}^{-i}(k,k) = \mathrm{Tor}_{i-1}^{R/\boldsymbol{x}R}(k,k)$ for any $i > 0$, and making this substitution in the previous formula yields the desired result (A.10.4).

APPENDIX B

Gorenstein Noether algebras

One of the main results in Buchweitz's manuscript, in his own assessment, is the equivalence of categories in Theorem 4.4.1. This is also borne out by its role in many of the subsequent developments in this area. The context for this result is that of an Iwanaga–Gorenstein ring, that is to say, a ring that is noetherian and of finite injective dimension, on both sides. The statement asserts equivalences between three different triangulated categories naturally associated with such a ring. In this appendix we explain how these various equivalences fit into a broader context, especially for rings of infinite Krull dimension. For developments taking Theorem 4.4.1 in different directions, see Appendix C and also, for example, the work of Keller and Vossieck [**86**]; Beligiannnis [**26**]; and Iyama and Yang [**72**].

B.1 We begin with the following result that combines contributions of Bass [**23**] and Goto [**61**].

THEOREM B.1.1. *Let S be a commutative noetherian ring. The following conditions are equivalent.*

(1) *The ring $S_{\mathfrak{p}}$ has finite injective dimension for each $\mathfrak{p}$ in* $\operatorname{Spec} S$.
(2) *Each M in* $\operatorname{mod}(S)$ *satisfies* $\operatorname{Ext}^i_S(M,S)=0$ *for* $i \gg 0$.
(3) *Each M in* $\operatorname{mod}(S)$ *has a syzygy module that is G-projective.* □

The modules in condition (3) are defined, in a broader context, as follows:

DEFINITION B.1.2. Let S be a ring, noetherian on both sides. A finitely generated (right) S-module M is said to be *G-projective* (short for *Gorenstein projective*) if it satisfies the following conditions:

(1) $\operatorname{Ext}^i_S(M,S)=0$ for $i \geq 1$,
(2) $\operatorname{Ext}^i_{S^{\mathrm{op}}}(M^*,S)=0$ for $i\geq 1$, and
(3) $M \to M^{**}$ is bijective.

Auslander and Bridger [**AB69**] introduced the notion under the name *modules of G-dimension zero* and provided several characterizations; we have selected the one in [**AB69**, (3.8)]. In the literature, these modules are also called *totally reflexive*. We refer the reader to [**17**] for a discussion on the various perspectives behind these names. See also Theorem B.1.7 and the paragraph ensuing that.

When S is Iwanaga–Gorenstein, condition (1) means M is MCM, and it implies conditions (2) and (3) by Lemma 4.2.2.

The full subcategory of $\operatorname{mod}(S)$ consisting of G-projective modules is denoted $\operatorname{GP}(S)$. It is a Frobenius exact subcategory of $\operatorname{mod}(S)$. We write $\underline{\operatorname{GP}}(S)$ for the associated stable category; this is a triangulated category [**Hap85**, Chapter I.2]. The relation to Buchweitz's work is apparent from a reformulation of conditions (1)–(3): A complex A of finitely generated projective S-modules such that both

A and $\operatorname{Hom}_S(A,S)$ are acyclic is said to be *totally acyclic*. A finitely generated S-module M is G-projective if and only if there is a totally acyclic complex A and an isomorphism $\Omega_0(A) \cong M$; in this situation A is called a *totally acyclic resolution of M*.

Let $\mathrm{K}_{\mathrm{tac}}(\operatorname{proj} S)$ denote the full subcategory of $\mathrm{K}(S)$ consisting of totally acyclic complexes of projectives, and let

$$\Omega_0\colon \mathrm{K}_{\mathrm{tac}}(\operatorname{proj} S) \longrightarrow \underline{\mathrm{GP}}(S)\,,$$

be the functor that associates to a complex its syzygy module in degree zero.

THEOREM B.1.3. *The functor Ω_0 is an equivalence of triangulated categories.*

A proof of this result can be extracted from that of Theorem 4.4.1; see (B.2.2) for a sketch. Inspired by Theorem B.1.1 we introduce the following definition.

DEFINITION B.1.4. Let S be a ring, noetherian on both sides. We say S is *Gorenstein* if for each finitely generated S-module L one has

$$\operatorname{Ext}^i_S(L,S) = 0 \quad \text{for all } i \gg 0$$

and the same for modules over S^{op}.

Evidently an Iwanaga–Gorenstein ring is Gorenstein in this sense, but the converse does not hold even for commutative rings; see Theorem B.1.1 and Nagata's example of a regular (and hence Gorenstein) ring of infinite Krull dimension [**103**, Appendix A1].

DEFINITION B.1.5. We say that a ring S is a *Noether R-algebra* if it is an algebra over a commutative noetherian ring R, and finite as an R-module; in particular, the ring S is noetherian on both sides. The ring R is omitted from the notation when it is not pertinent to the discussion. A ring S is a Noether algebra precisely when it is finite as a module over $\mathrm{Z}(S)$ and the latter ring is noetherian; see [**98**, Theorem 3.7].

The result below, proved in (B.3.5), is some validation of Definition B.1.4; confer Theorem B.1.1.

THEOREM B.1.6. *Let S be a Noether R-algebra. The following statements are equivalent.*

(1) *The ring S is Gorenstein, in the sense of* B.1.4.
(2) *Each M in* $\operatorname{mod}(S)$ *or* $\operatorname{mod}(S^{\mathrm{op}})$ *has a syzygy module that is G-projective.*

When they hold $S_{\mathfrak{p}}$ is Iwanaga–Gorenstein for each $\mathfrak{p}$ in $\operatorname{Spec} R$.

We deduce (1)$\Rightarrow$(2) from the more general result below.

THEOREM B.1.7. *Let S be a Gorenstein Noether algebra.*

(1) *Any acyclic complex of finitely generated projective S-modules is totally acyclic.*
(2) *A finitely generated S-module M is G-projective if and only if*

$$\operatorname{Ext}^i_S(M,S) = 0 \quad \textit{for } i \geq 1.$$

This result, proved in (B.3.4), and Theorem B.1.8 below bring us back to Buchweitz's work, for it identifies the G-projective modules as analogues of maximal Cohen–Macaulay modules.

THEOREM B.1.8. *Let S be a Gorenstein Noether algebra. The natural embedding* $\operatorname{mod}(S) \to \mathrm{D^b}(S)$ *induces an equivalence of triangulated categories:*

$$\iota_S \colon \underline{\mathrm{GP}}(S) \longrightarrow \underline{\mathrm{D}}^{\mathrm{b}}(S)\,.$$

See (B.4.4) for a proof. Observe that, at least for Noether algebras, Theorems B.1.8 and B.1.3 contain Theorem 4.4.1. In particular, we get an extension of the latter to commutative Gorenstein rings of infinite Krull dimension.

B.2 Morphisms between totally acyclic complexes. In what follows, given a complex X of S-modules we consider the inverse system $\sigma_{\leqslant i}X$ whose transition maps we denote by

$$\sigma \colon \sigma_{\leqslant i}X \to \sigma_{\leqslant i-1}X\,.$$

The result below is well-known; details are given for ease of reference. In what follows we write $\operatorname{Hom}_{\mathrm{K}(S)}(-,-)$, or just $\operatorname{Hom}_K(-,-)$, for the morphisms in the homotopy category of complexes of S-modules, that is to say, morphisms of complexes modulo the homotopy relation.

LEMMA B.2.1. *Let S be a right noetherian ring, and A, B complexes of finitely generated projective S-modules. When A is totally acyclic, for each i the natural map*

$$\operatorname{Hom}_{\mathrm{K}(S)}(A,B) \to \operatorname{Hom}_{\mathrm{K}(S)}(A,\sigma_{\leqslant i}B)$$

is an isomorphism; hence so is the map $\operatorname{Hom}_{\mathrm{K}(S)}(A,B) \to \lim_i \operatorname{Hom}_{\mathrm{K}(S)}(A,\sigma_{\leqslant i}B)$.

PROOF. From the degree-wise split exact sequence

$$0 \to \mathrm{T}^{-i} B^i \to \sigma_{\leqslant i}B \xrightarrow{\ \sigma\ } \sigma_{\leqslant i-1}B \to 0\,,$$

we obtain a long exact sequence

$$\cdots \to \operatorname{Hom}_K(A,\mathrm{T}^{-i} B^i) \to \operatorname{Hom}_K(A,\sigma_{\leqslant i}B) \xrightarrow{\ \sigma_*\ } \operatorname{Hom}_K(A,\sigma_{\leqslant i-1}B) \to \cdots$$

of abelian groups. Since A is totally acyclic and each B^i is a finitely generated projective module we deduce that $\operatorname{Hom}_K(A,\mathrm{T}^{-i} B^i)=0$, and so σ_* is an isomorphism. Knowing this, it suffices to prove the *second* statement of the lemma.

There is another obvious, degree-wise split, short exact sequence

$$0 \longrightarrow B \longrightarrow \prod \sigma_{\leqslant i}B \xrightarrow{\ \mathrm{id}-\sigma\ } \prod \sigma_{\leqslant i}B \longrightarrow 0$$

of chain complexes. From this we obtain a long exact sequence

$$\cdots \longrightarrow \operatorname{Hom}_K(A,B) \longrightarrow \prod \operatorname{Hom}_K(A,\sigma_{\leqslant i}B) \xrightarrow{\ \mathrm{id}-\sigma_*\ } \prod \operatorname{Hom}_K(A,\sigma_{\leqslant i}B) \longrightarrow \cdots$$

of abelian groups. Since the σ_* are isomorphisms the right-hand map $\mathrm{id}-\sigma_*$ is surjective. Hence this long exact sequence naturally identifies $\operatorname{Hom}_K(A,B)$ with

$$\lim_i \operatorname{Hom}_K(A,\sigma_{\leqslant i}B) = \ker\Big(\prod \operatorname{Hom}_K(A,\sigma_{\leqslant i}B) \xrightarrow{\ \mathrm{id}-\sigma_*\ } \prod \operatorname{Hom}_K(A,\sigma_{\leqslant i}B)\Big)$$

This completes the proof. □

(B.2.2) *Proof of Theorem* B.1.3. It is clear from the definition of G-projectives that Ω_0 is surjective on objects. Let A, B be totally acyclic complexes of projective S-modules. There is a chain of natural isomorphisms

$$\begin{aligned}\operatorname{Hom}_K(A,B) &\xrightarrow{\cong} \operatorname{Hom}_K(A,\sigma_{\leqslant 0}B)\\ &\xrightarrow{\cong} \operatorname{Hom}_K(A,\Omega_0(B))\\ &\xrightarrow{\cong} \underline{\operatorname{Hom}}_S(\Omega_0(A),\Omega_0(B))\end{aligned}$$

where the first one is induced by the surjection $B \to \sigma_{\leqslant 0}B$; it is an isomorphism by Lemma B.2.1. The second map is induced by the surjection $\sigma_{\leqslant 0}B \to \Omega_0(B)$; this is a quasi-isomorphism as B is acyclic. Since A is a complex of projectives, $\operatorname{Hom}_S(A,-)$ preserves quasi-isomorphisms between bounded above complexes; thus the second map above is an isomorphism. Any morphism of complexes $A \to \Omega_0(B)$ factors through $\Omega_0(A)$; using the fact that $\operatorname{Ext}^1_S(\Omega_0(A),S) = 0$ it is a routine exercise to verify that the induced map, which is the third one above, is an isomorphism. Thus the functor Ω_0 is fully faithful. Since it is also surjective on objects, it follows that it is an equivalence of categories. □

B.3 Homological dimensions over Noether algebras. Throughout this section S will be a Noether algebra. The finiteness of the projective dimension of a finitely generated S-module M is characterized by the vanishing of $\operatorname{Ext}^i_S(M,-)$ for $i \gg 0$; see [**51**, Appendix 2]. The injective dimension of a complex is as in [**11**]; for modules it is the usual notion.

LEMMA B.3.1. *Let S be a semi-local ring and M a complex in $\mathrm{D^b}(S)$. If S is a Noether algebra, then the following equality holds:*

$$\operatorname{injdim}_S M = \sup\{j \in \mathbb{Z} \mid \operatorname{Ext}^j_S(k,M) \neq 0,\ \ k \text{ a simple } S\text{-module}\}\,.$$

PROOF. Set $R := \mathrm{Z}(S)$. Since S is semi-local and finite over R, the ring R is semi-local. Let $\mathfrak{q}$ denote the Jacobson radical of R and $\mathfrak{a}$ that of S. Since S is semi-local there exists a natural number q such that $\mathfrak{a}^q \subseteq \mathfrak{q}S \subseteq \mathfrak{a}$ hold; see [**93**, 20.6]. Thus S has only finitely many simple S-modules, and an S-module has finite length if and only if it has finite length over R.

Let s be the number on the right-hand side of the desired equality. We have

$$\operatorname{injdim}_S M = \sup\{j \in \mathbb{Z} \mid \operatorname{Ext}^j_S(L,M) \neq 0, L \text{ a finite } S\text{-module}\} \geq s\,,$$

with equality from [**11**, (2.5.I)]. Thus, it suffices to show that if s is finite, then

$$\operatorname{Ext}^j_S(L,M) = 0 \quad \text{for all} \quad j > s\,. \tag{B.3.2}$$

We induce on $d := \dim_R L$, the Krull dimension of the R-module L. As L is a finite R-module, $d = 0$ means $\operatorname{length}_S L = l < \infty$. If $l = 1$, then L is simple and (B.3.2) holds by the definition of s. For $l > 1$ pick a non-zero, proper S-submodule L' of L. Now (B.3.2) results by induction on l from the exact sequence

$$\operatorname{Ext}^j_S(L/L',M) \longrightarrow \operatorname{Ext}^j_S(L,M) \longrightarrow \operatorname{Ext}^j_S(L',M)\,.$$

Assume next $d \geq 1$ and the formula holds for all S-modules of dimension smaller than d. The set $L'' = \bigcup_h \left(0 :_L \mathfrak{q}^h\right)$ is an S-submodule of L; it has finite length over R, and hence over S. For the S-module $\overline{L} := L/L''$ the exact sequence

$$\operatorname{Ext}^j_S(\overline{L},M) \longrightarrow \operatorname{Ext}^j_S(L,M) \longrightarrow \operatorname{Ext}^j_S(L'',M)$$

shows that it suffices to prove $\mathrm{Ext}^j_S(\overline{L}, M) = 0$ for all $j > s$. The R-module $\overline{L}$ has positive depth. Choosing in $\mathfrak{q}$ a non-zero-divisor x on $\overline{L}$, we get an exact sequence

$$0 \longrightarrow \overline{L} \xrightarrow{x} \overline{L} \longrightarrow \overline{L}/x\overline{L} \longrightarrow 0$$

of S-modules, and hence an exact sequence of R-modules

$$\mathrm{Ext}^j_S(\overline{L}, M) \xrightarrow{x} \mathrm{Ext}^j_S(\overline{L}, M) \longrightarrow \mathrm{Ext}^{j+1}_S(\overline{L}/x\overline{L}, M)\,.$$

Since $\dim_S(\overline{L}/x\overline{L}) = d-1$, the module on the right is equal to 0 by the induction hypothesis. This gives $\mathrm{Ext}^j_S(\overline{L}, M) = x\,\mathrm{Ext}^j_S(\overline{L}, M)$. But $\mathrm{Ext}^j_S(\overline{L}, M)$ is a finite R-module, so Nakayama's Lemma implies that it is equal to 0, as desired. □

A standard localization argument now the yields the following result.

COROLLARY B.3.3. *Let S be a Noether R-algebra and $L \in \mathrm{D}^{\mathrm{b}}(S)$. If one has*

$$\mathrm{Ext}^i_S(M, L) = 0 \quad \textit{for each } M \in \mathrm{mod}(S) \textit{ and } i \gg 0,$$

then $\operatorname{injdim}_{S_{\mathfrak{p}}} L_{\mathfrak{p}} < \infty$ *for each* $\mathfrak{p}$ *in* $\operatorname{Spec} R$. □

(B.3.4) *Proof of Theorem* B.1.7. Recall that S is a Gorenstein Noether algebra; set $R := \mathrm{Z}(S)$. The crucial point in the proof is that both statements concern vanishing of certain cohomology modules, and this is a local property: An R-module, say E, is zero if and only if $E_{\mathfrak{p}} = 0$ for each $\mathfrak{p}$ in $\operatorname{Spec} R$. By Corollary B.3.3 the ring $S_{\mathfrak{p}}$ has finite injective dimension, so one has only to verify the statements when the injective dimension of S is finite, in which case they are already known: For (1) see Remark A.5.2 and for (2) see Lemma 4.2.2.

This completes the proof of Theorem B.1.7. □

(B.3.5) *Proof of Theorem* B.1.6. Recall that S is a Noether algebra.

(1)⇒(2) Let M be a finitely generated S-module. Since S is assumed to be Gorenstein, there exists an integer $d \geq 1$ such that $\mathrm{Ext}^i_S(M, S) = 0$ for $i \geq d+1$. Thus for $j \geq 1$ one has

$$\mathrm{Ext}^j_S(\Omega_{-d} M, S) \cong \mathrm{Ext}^{j+d}_S(M, S) = 0\,.$$

Then applying Theorem B.1.7 we deduce that $\Omega_{-d}M$ is G-projective. Since the Gorenstein condition is symmetric, the same conclusion holds also for S^{op}-modules.

(2)⇒(1) Fix M in $\mathrm{mod}(S)$ or $\mathrm{mod}(S^{\mathrm{op}})$. The hypothesis is that $\Omega_{-d}M$ is G-projective for some $d \geq 0$. In particular for $i \geq d+1$ one gets

$$\mathrm{Ext}^i_S(M, S) \cong \mathrm{Ext}^{i-d}_S(\Omega_{-d} M, S) = 0\,.$$

The last assertion in the statement is the special case $L = S$ of Corollary B.3.3.

This completes the proof of Theorem B.1.6. □

B.4 Verdier's criterion. Verdier [**124**, Proposition 2.3.3] proved the following result for computing morphisms in quotient categories:

THEOREM B.4.1. *Let A be a triangulated category and $\mathsf{B} \subseteq \mathsf{A}$ a thick subcategory. Fix an object X in A. Then the natural transformation of functors*

$$\mathrm{Hom}_{\mathsf{A}}(X, -) \longrightarrow \mathrm{Hom}_{\mathsf{A}/\mathsf{B}}(X, -)$$

is an isomorphism if and only if $\mathrm{Hom}_{\mathsf{A}}(X, \mathsf{B}) = 0$. □

In what follows we need the following version of this result, where there may be no uniform k such that $\operatorname{Hom}_{\mathsf{A}}(X_k, \mathsf{B}) = 0$. The proof shows that the inverse system could just as well be any cofiltered diagram in A.

LEMMA B.4.2. *Let* A *be a triangulated category and* $\mathsf{B} \subseteq \mathsf{A}$ *a thick subcategory. Given a system of maps* $X_0 \leftarrow X_1 \leftarrow \cdots$ *in* A*, the induced natural transformation of functors*

$$\operatorname{colim} \operatorname{Hom}_{\mathsf{A}}(X_k, -) \longrightarrow \operatorname{colim} \operatorname{Hom}_{\mathsf{A}/\mathsf{B}}(X_k, -)$$

is an isomorphism if and only if $\operatorname{colim} \operatorname{Hom}_{\mathsf{A}}(X_k, \mathsf{B}) = 0$.

PROOF. The only if direction is trivial.

Assume $\operatorname{colim} \operatorname{Hom}_{\mathsf{A}}(X_k, \mathsf{B}) = 0$. This means that $\operatorname{colim} \operatorname{Hom}_{\mathsf{A}}(X_k, -)$ descends to A/B. To be more precise, $\operatorname{colim} \operatorname{Hom}_{\mathsf{A}}(X_k, -) \colon \mathsf{A} \to \mathrm{Ab}$ is a homological functor (by exactness of filtered colimits) which vanishes on B, and therefore there is a functor $F \colon \mathsf{A}/\mathsf{B} \to \mathrm{Ab}$ and a natural isomorphism $\alpha \colon \operatorname{colim} \operatorname{Hom}_{\mathsf{A}}(X_k, -) \to F\pi$ where $\pi \colon \mathsf{A} \to \mathsf{A}/\mathsf{B}$ is the natural projection. Note that

$$\operatorname{colim} \operatorname{Hom}_{\mathsf{A}}(X_k, -) \to \operatorname{colim} \operatorname{Hom}_{\mathsf{A}/\mathsf{B}}(X_k, -)$$

factors as

$$\operatorname{colim} \operatorname{Hom}_{\mathsf{A}}(X_k, -) \xrightarrow{\ \alpha\ } F\pi \xrightarrow{\ \beta\ } \operatorname{colim} \operatorname{Hom}_{\mathsf{A}/\mathsf{B}}(X_k, -)\,.$$

Indeed, since A/B has the same objects as A we have to define β on objects by

$$F(Y) \xrightarrow{\ \alpha^{-1}\ } \operatorname{colim} \operatorname{Hom}_{\mathsf{A}}(X_k, Y) \longrightarrow \operatorname{colim} \operatorname{Hom}_{\mathsf{A}/\mathsf{B}}(X_k, Y)\,.$$

And because morphisms in A/B are fractions of morphisms from A, naturality of β follows from its naturality when restricted to A.

Now we show, by a Yoneda style argument, that the existence of this factorization formally implies that $\operatorname{colim} \operatorname{Hom}_{\mathsf{A}}(X_k, Y) \to \operatorname{colim} \operatorname{Hom}_{\mathsf{A}/\mathsf{B}}(X_k, Y)$ is bijective for any Y in A.

Injectivity. If f is in the kernel of $\operatorname{colim} \operatorname{Hom}_{\mathsf{A}}(X_k, Y) \to \operatorname{colim} \operatorname{Hom}_{\mathsf{A}/\mathsf{B}}(X_k, Y)$ then f is represented a morphism $f_i \colon X_i \to Y$, for some i, which vanishes in A/B, after applying π. From the naturality square for α

$$\begin{array}{ccc}
\operatorname{colim} \operatorname{Hom}_{\mathsf{A}}(X_k, X_i) & \xrightarrow{\ \cong\ } & F(X_i) \\
{\scriptstyle (f_i)_*}\downarrow & & \downarrow{\scriptstyle F(\pi(f_i))=0} \\
\operatorname{colim} \operatorname{Hom}_{\mathsf{A}}(X_k, Y) & \xrightarrow{\ \cong\ } & F(Y)
\end{array}$$

it follows that $f = (f_i)_*(\mathrm{id}_{X_i}) = 0$.

Surjectivity. An element g in $\operatorname{colim} \operatorname{Hom}_{\mathsf{A}/\mathsf{B}}(X_k, Y)$ is represented by a morphism $g_i \colon X_i \to Y$ in A/B for some i. Consider the naturality square for β

$$\begin{array}{ccc}
F(X_i) & \longrightarrow & \operatorname{colim} \operatorname{Hom}_{\mathsf{A}/\mathsf{B}}(X_k, X_i) \\
{\scriptstyle F(g_i)}\downarrow & & \downarrow{\scriptstyle (g_i)_*} \\
F(Y) & \longrightarrow & \operatorname{colim} \operatorname{Hom}_{\mathsf{A}/\mathsf{B}}(X_k, Y)\,.
\end{array}$$

Since id_{X_i} in A/B is the image of id_{X_i} in A under $\beta\alpha$, it follows that

$$\beta(F(g_i)(\alpha(\mathrm{id}_{X_i}))) = g$$

is in the image of β, and hence in the image of $\beta\alpha$. $\square$

The criterion will be used in the following form:

LEMMA B.4.3. *Let S be a right noetherian ring, A in $\mathrm{K_{tac}}(\operatorname{proj} S)$ and Y a bounded above complex of finitely generated S-modules with bounded cohomology. The natural map*

$$\operatorname{Hom}_{\mathrm{K}(S)}(A,Y) \longrightarrow \operatorname{Hom}_{\underline{\mathrm{D}}^{\mathrm{b}}(S)}(\underline{\sigma}_{\leqslant}A,Y)$$

is an isomorphism.

PROOF. Since $\sigma_{\leqslant k}A$ is a bounded above complex of projectives, one has

$$\operatorname{Hom}_{\mathrm{D^b}(S)}(\sigma_{\leqslant k}A,-) = \operatorname{Hom}_{\mathrm{K}(S)}(\sigma_{\leqslant k}A,-)\,.$$

The argument given in part (v) of the proof of Theorem 4.4.1 shows that the functor

$$\operatorname{colim}\operatorname{Hom}_{\mathrm{D^b}(S)}(\sigma_{\leqslant k}A,-)$$

vanishes on $\mathrm{D}^{\mathrm{b}}_{\mathrm{perf}}(S)$. The induced functor

$$\operatorname{colim}\operatorname{Hom}_{\underline{\mathrm{D}}^{\mathrm{b}}(S)}(\underline{\sigma}_{\leqslant k}A,-)$$

is representable by $\underline{\sigma}_{\leqslant}A$, because the transition maps become isomorphisms in $\underline{\mathrm{D}}^{\mathrm{b}}(S)$. According to Lemma B.4.2 the natural transformation

$$\operatorname{colim}\operatorname{Hom}_{\mathrm{D^b}(S)}(\sigma_{\leqslant k}A,-) \longrightarrow \operatorname{Hom}_{\underline{\mathrm{D}}^{\mathrm{b}}(S)}(\underline{\sigma}_{\leqslant}A,-)$$

is an isomorphism. Because Y is bounded above the map

$$\operatorname{Hom}_{\mathrm{D^b}(S)}(\sigma_{\leqslant i}A,Y) = \operatorname{Hom}_{\mathrm{K}(S)}(\sigma_{\leqslant i}A,Y) \longrightarrow \operatorname{Hom}_{\mathrm{K}(S)}(A,Y)$$

is an isomorphism for large enough i, so there are canonical isomorphisms

$$\operatorname{Hom}_{\mathrm{K}(S)}(A,Y) \cong \operatorname{colim}\operatorname{Hom}_{\mathrm{D^b}(S)}(\sigma_{\leqslant k}A,Y) \cong \operatorname{Hom}_{\underline{\mathrm{D}}^{\mathrm{b}}(S)}(\underline{\sigma}_{\leqslant}A,Y)\,. \qquad \square$$

(B.4.4) *Proof of Theorem* B.1.8. The hypothesis is that S is a Gorenstein Noether algebra. The plan is to verify that the composition

$$\mathrm{K_{tac}}(\operatorname{proj} S) \xrightarrow{\ \Omega_0\ } \underline{\mathrm{GP}}(S) \xrightarrow{\ \iota_S\ } \underline{\mathrm{D}}^{\mathrm{b}}(S)$$

is an equivalence of categories. This yields the desired result for we already know that the functor Ω_0 is an equivalence, by Theorem B.1.3.

The functor $\iota_S \circ \Omega_0$ assigns to each $A \in \mathrm{K_{tac}}(\operatorname{proj} S)$ the object $\underline{\sigma}_{\leqslant}A$. This functor is fully faithful because for A, B in $\mathrm{K_{tac}}(\operatorname{proj} S)$ one has isomorphisms

$$\operatorname{Hom}_{\mathrm{K}(S)}(A,B) \xrightarrow{\ \cong\ } \operatorname{Hom}_{\mathrm{K}(S)}(A,\sigma_{\leqslant i}B) \xrightarrow{\ \cong\ } \operatorname{Hom}_{\underline{\mathrm{D}}^{\mathrm{b}}(S)}(\underline{\sigma}_{\leqslant}A,\underline{\sigma}_{\leqslant}B)$$

for any i. The first isomorphism is by Lemma B.2.1 and the second one is by Lemma B.4.3. It remains to verify that $\iota_S\circ\Omega_0$ is essentially surjective; equivalently, that ι_S is essentially surjective; equivalently, that any complex X is isomorphic in $\underline{\mathrm{D}}^{\mathrm{b}}(S)$ to a G-projective module.

We can assume X is a bounded above complex of finitely generated projective modules. Then $X \cong \sigma_{\leqslant i}X$ for some i. Choosing $i \ll 0$ we can assume $\sigma_{\leqslant i}X$ is isomorphic to $M[-i]$, where $M := \Omega_{-i}X$. From Theorem B.1.6 we get that $\Omega_{-d}M$ is G-projective for an integer $d \geq 0$. Thus in $\underline{\mathrm{D}}^{\mathrm{b}}(S)$ one has

$$X \cong M[-i] \cong \Omega_{-d}M[-i] \cong \Omega_{i-d}M\,.$$

Since $\Omega_{i-d}M$ is G-projective, this completes the proof. $\square$

APPENDIX C

Subsequent developments

This chapter is a survey of some developments related to Buchweitz's manuscript in the thirty-five years since it was written. The choice of topics and the level of detail is circumscribed by our interests, and more so by our expertise. Notation and terminology are as in Chapters 1–10, except that we write "Iwanaga–Gorenstein" in place of "strongly Gorenstein".

C.1 Koszul duality and the BGG correspondence. In Chapter 9 of this text the Bernšteĭn–Gelfand–Gelfand correspondence is placed in context and extended to the case of projective superspaces. Here we summarise what forms the BGG correspondence has taken in the decades since.

Let $S := k[x_1, \ldots, x_n]$ and $\Lambda := \bigwedge(y_1, \ldots, y_n)$ be a polynomial algebra and an exterior algebra, each *standard graded*, meaning that they are graded and generated as k-algebras by their degree one components. The classical BGG correspondence describes an equivalence of triangulated categories

$$\underline{\mathrm{MCM}}_{\mathrm{gr}}(\Lambda) \cong \mathrm{D}^{\mathrm{b}}(\mathrm{Proj}\text{-}S)\,,$$

where $\mathrm{Proj}\text{-}S := \mathrm{mod}(S)/\mathrm{art}(S)$ is the abelian quotient of the category of finitely generated graded modules by those of finite total length, as in Chapter 9. At its core, this correspondence exists because S and Λ are Koszul dual algebras, in the sense that

$$S \cong \mathrm{Ext}_{\Lambda}(k,k) \quad \text{and} \quad \Lambda \cong \mathrm{Ext}_{S}(k,k)\,. \tag{C.1.1}$$

Therefore we begin by making this into a definition: Two k-algebras R and E are *Koszul dual* to each other if both are standard graded and

$$R \cong \mathrm{Ext}_{E}(k,k) \quad \text{and} \quad E \cong \mathrm{Ext}_{R}(k,k)\,, \tag{C.1.2}$$

where the Ext-algebras are given their natural cohomological grading. Any algebra which is part of a Koszul dual pair is called a *Koszul algebra*.

The algebras $\mathbf{U}(V,W)$ and $\mathbf{U}(W^*,V^*)$, which are the subject of Chapter 9, are Koszul dual to each other. This generalization of (C.1.1) fuels the BGG correspondence for sheaves on projective superspaces.

The next theorem, proven by Buchweitz [**BEH87**, App.] at around the time this manuscript was prepared, motivates the non-commutative framework built up in the preceding chapters. Indeed, this work was in part designed to serve as a general foundation for results such as the one below (cf. the original preface).

THEOREM C.1.3. *Assume that R and E are graded Gorenstein two-sided noetherian algebras which are Koszul dual to each other. There are equivalences*

$$\underline{\mathrm{MCM}}_{\mathrm{gr}}(R) \cong \mathrm{D}^{\mathrm{b}}(\mathrm{Proj}\text{-}E) \quad \text{and} \quad \underline{\mathrm{MCM}}_{\mathrm{gr}}(E) \cong \mathrm{D}^{\mathrm{b}}(\mathrm{Proj}\text{-}R)\,.$$

Here $\operatorname{Proj}\text{-}E := \operatorname{mod}(E)/\operatorname{art}(E)$, defined as above, should be interpreted through the lens of non-commutative projective geometry [**2**].

The motivating example for Theorem C.1.3 is the case when

$$R := k[x_1, \ldots, x_n]/(q_1, \ldots, q_c)$$

is complete intersection defined by quadrics—that is to say, the polynomial ring is standard graded and each q_i has degree 2, and $q_1, \ldots, q_c$ form a regular sequence.

The computation of $E := \operatorname{Ext}_R(k,k)$—and the proof that E and R form a Koszul dual pair of graded Gorenstein noetherian algebras—is given in [**BEH87**, App.]. To give a quick overview: The defining relations $q_1, \ldots, q_c$ yield naturally a graded quadratic form $Q\colon S^n \to S$ over the polynomial ring $S := k[z_1, \ldots, z_c]$, where each z_i is assigned degree 2. The Ext-algebra E is then isomorphic to the graded Clifford-algebra of this quadratic form:

$$E \cong \operatorname{Cliff}_S(Q)\,.$$

In order to understand maximal Cohen–Macaulay modules over R, Buchweitz also describes the category $\operatorname{Proj}\text{-}\operatorname{Cliff}_S(Q)$, which is equivalent to $\underline{\operatorname{MCM}}_{\mathrm{gr}}(R)$ by Theorem C.1.3. Indeed, one can sheafify $\operatorname{Cliff}_S(Q)$ to obtain a sheaf of (ordinary, finite dimensional) Clifford-algebras on $\mathbb{P}_k^{c-1} = \operatorname{Proj}(S)$, and thereby describe the objects of $\operatorname{Proj}\text{-}\operatorname{Cliff}_S(Q)$ as sheaves of Clifford modules on $\mathbb{P}_k^{c-1}$; see [**BEH87**, App.].

Jumping forwards in time, Beĭlinson, Ginzburg and Soergel [**24**] describe equivalences of derived categories arising from a Koszul dual pair of algebras in much greater generality. The derived Morita theory developed by Keller [**81**] yields derived equivalences in even more situations; in this context one does not even need a Koszul algebra, as long as one is willing to work with *differential graded algebras*. In this direction, Greenlees and Stevenson [**63**] have developed a framework for producing BGG style equivalences based on examples from algebraic topology. However, the equivalences appearing in these works do not concern bounded derived categories in the usual sense—in fact the algebras involved are rarely noetherian—and there is no clear analogue for the derived category of sheaves on some (possibly non-commutative) space.

For geometric results in the style of the original BGG correspondence, a theorem of Bøgvad and Halperin [**30**] provides a serious obstacle: If R is a commutative noetherian graded algebra over k, then $\operatorname{Ext}_R(k,k)$ is noetherian if and only if R is a complete intersection ring. In other words, Theorem C.1.3 applies to *no other examples* in commutative algebra, beyond the quadratic complete intersection rings.

Nonetheless Gélinas [**59**] proved a substantial generalization of Theorem C.1.3. One of the key observations of [**59**] is that there are many examples of commutative Koszul Gorenstein rings R such that the Koszul dual ring E is *graded coherent*. By definition a graded ring E is right graded coherent if all of its finitely generated homogeneous right ideals are finitely presented. Left graded coherent rings are defined similarly, and graded coherent rings must have both properties. See the work of Cohen [**47**] and Minamoto [**100**] for more information. To produce examples of Koszul algebras with graded coherent Koszul duals, Gélinas considers the class of absolutely Koszul algebras introduced by Herzog and Iyengar [**67**].

Let R be a standard graded k-algebra and let F be a complex of graded free R-modules. For each n consider the subcomplex $F_{(\geqslant n)}$ generated by the subspaces F_j^i with $i + j \geq n$ (here i is the cohomological index and j is the internal degree).

Using these we set

$$\operatorname{lin}^R_n(F) := F_{(\geqslant n)}/F_{(\geqslant n+1)} \quad \text{and} \quad \operatorname{lin}^R(F) := \bigoplus_n \operatorname{lin}^R_n(F)\,.$$

If one represents the differential of F as a sequence of matrices, then $\operatorname{lin}^R(F)$ is the complex obtained from F by setting all non-linear entries in these matrices to zero.

We say that a right R-module M has *finite linearity defect* if there is a resolution

$$\cdots \to F^{-n} \to \cdots \to F^{-1} \to F^0 \to M \to 0\,,$$

such that $\operatorname{lin}^R(F)$ has finitely many non-zero homology groups: $\mathrm{H}^{-i}(\operatorname{lin}^R F) = 0$ for $i \gg 0$. By definition R is right *absolutely Koszul* if it is right coherent and all of its finitely presented right R-modules have finite linearity defect. Left absolutely Koszul algebras are defined similarly; if R is commutative the two conditions are the same. This is stronger than the Koszul property. There are many examples of absolutely Koszul algebras, but also many examples of algebras that are Koszul but not absolutely Koszul; see [**48**, **67**, **74**].

THEOREM C.1.4 (Gélinas [**59**]). *Assume that R is right absolutely Koszul Gorenstein with Koszul dual E. Then E is right graded coherent and there is an equivalence of triangulated categories*

$$\underline{\mathrm{MCM}}_{\mathrm{gr}}(R) \cong \mathrm{D}^{\mathrm{b}}(\operatorname{Proj}\text{-}E)\,.$$

In this context $\operatorname{Proj}\text{-}E$ is defined to be the abelian quotient category of the category of *finitely presented* graded modules by those of finite total length. The theorem is proven by constructing a t-structure on $\underline{\mathrm{MCM}}_{\mathrm{gr}}(R)$, called the *linear t-structure*, which realises $\operatorname{Proj}\text{-}E$ as its heart.

In commutative algebra there are many rings to which the theorem applies, and thereby many new instances of the BGG correspondence. Indeed, R is an absolutely Koszul Gorenstein ring if:

- R is a quadratic complete intersection;
- R is a short Gorenstein ring [**15**], [**59**, 6.1.1];
- R is Koszul Gorenstein and generalized Golod; this includes the previous two examples, any Gorenstein ring that has the Backelin–Roos property [**67**, Theorem 5.9], and any Gorenstein ring of embedding codimension at most 4 [**16**, Proposition 6.3];
- R is the homogeneous coordinate ring of a smooth projective variety X, defined over an algebraically closed field, of dimension $n \geq 1$, degree $d \geq 4$, and embedded in $\mathbb{P}^{d+n-2}_k$; when $n = 1$ the curve X is elliptic; see [**59**, 6.1.5];
- R is a Veronese subring of a polynomial satisfying certain conditions, see [**48**].

For more examples see [**59**, §6.1]. In several of these examples Gélinas describes explicitly both sides of the BGG correspondence of Theorem C.1.4.

Gélinas [**59**, §6.2] used these ideas to disprove coherence conjectures of Bondal [**112**] and Minamoto [**100**]. His counterexamples rely on the existence of Koszul Gorenstein ring that are not absolutely Koszul. Such rings were first constructed by Jacobsson [**78**, Corollary 2.3]; see also the work of Roos [**116**].

C.2 Knörrer periodicity and stable equivalences. It has become an important theme in representation theory to understand what properties are shared by Iwanaga–Gorenstein rings A and B for which there is an equivalence of stable derived categories $\underline{\mathrm{D}}^{\mathrm{b}}(A) \simeq \underline{\mathrm{D}}^{\mathrm{b}}(B)$. The periodicity theorem of Knörrer [**Knö87**] is a fundamental, early example of this phenomenon.

Let k be an algebraically closed field. To a multivariable power series f without constant term in $k[\![x_1,\dots,x_n]\!]$ one can associate the *double branched cover* $f+y^2$ in $k[\![x_1,\dots,x_n,y]\!]$. Recast in the language developed by Buchweitz, it is proven in [**Knö87**], when the characteristic of k is not 2, that taking an iterated double branched cover results in a natural equivalence

$$\underline{\mathrm{D}}^{\mathrm{b}}\left(\frac{k[\![x_1,\dots,x_n]\!]}{(f)}\right) \simeq \underline{\mathrm{D}}^{\mathrm{b}}\left(\frac{k[\![x_1,\dots,x_n,y,z]\!]}{(f+y^2+z^2)}\right).$$

Knörrer uses this result to prove that all *simple* hypersurface singularities have only finitely many isomorphism classes of indecomposable MCM modules. The companion paper of Buchweitz, Greuel and Schreyer [**BGS87**] contains the converse: all non-simple hypersurface singularities have infinitely many non-isomorphic indecomposable MCM modules. See [**95**, Chapter 8] for more information.

Later, Solberg [**120**] reproved Knörrer's theorem in the case that the hypersurface $k[\![x_1,\dots,x_n]\!]/(f)$ has finite MCM representation type, and, by replacing the iterated double branched cover $f+y^2+z^2$ with $f+yz$, he removed all assumptions on the field k in this case.

The equivalences constructed by Knörrer share some structural similarities with the Bott periodicity isomorphisms from algebraic topology, which describe a 2-periodicity $\pi_n(U) \cong \pi_{n+2}(U)$ in the homotopy groups of the infinite unitary group U. In the case that $k=\mathbb{C}$, the two results are directly compatible according to the work of Brown; see [**33**] for a precise description of this compatibility. When $k=\mathbb{R}$, Brown establishes an 8-periodic analogue of Knörrer's result akin to the 8-periodicity in the homotopy groups of the infinite orthogonal group O.

After Buchweitz placed Knörrer's periodicity theorem into the framework of stable derived categories, it is natural to investigate the *stable Morita theory* of algebras in general. In contrast to Rickard's theorem on derived equivalences [**115**], it is not known whether two algebras with equivalent stable derived categories must admit an equivalence that is induced by tensoring with a bimodule. Thus, Broué [**32**] defined a *stable equivalence of Morita type* between finite dimensional self-injective algebras A and B to be a pair of modules M and N over $A^{\mathrm{op}}\otimes B$ and $B^{\mathrm{op}}\otimes A$ respectively, such that M and N are both finitely generated projective modules on either side, and such that $M\otimes_B N \simeq A$ in $\underline{\mathrm{D}}^{\mathrm{b}}(A^{\mathrm{ev}})$ and $N\otimes_A M\simeq B$ in $\underline{\mathrm{D}}^{\mathrm{b}}(B^{\mathrm{ev}})$. This context yields an exact equivalence

$$-\otimes_A M \;:\; \underline{\mathrm{D}}^{\mathrm{b}}(A) \overset{\simeq}{\rightleftarrows} \underline{\mathrm{D}}^{\mathrm{b}}(B) \;:\; -\otimes_B N\,.$$

Wang [**127**] generalized this to allow for algebras that are not finite dimensional. In either case, examples abound of non-trivial stable equivalences of Morita type. In particular, the work of Dyckerhoff and Murfet explains how one may view Knörrer's equivalences as examples of stable equivalences of Morita type; see [**53**, Section 13].

C.3 The work of Orlov. The stable derived category was rediscovered and extended to the geometric context by Orlov in [**107**], who called it the *singularity category*. For a separated, noetherian scheme X with enough locally free sheaves, Orlov defined the singularity category to be

$$\mathrm{D}_{\mathrm{sg}}(X) := \mathrm{D}^{\mathrm{b}}(\mathrm{Coh}X)/\mathrm{Perf}(X)\,,$$

the Verdier quotient of the bounded derived category of coherent sheaves on X by its subcategory of perfect complexes. To justify the name, Orlov proves that if $U \subseteq X$ is an open subscheme containing the singular locus of X, then the induced restriction functor

$$\mathrm{D_{sg}}(X) \to \mathrm{D_{sg}}(U)$$

is an equivalence [**107**, Proposition 1.14]; see also [**88**, Theorem 1.6].

In the affine case the singularity category coincides with the stable derived category introduced in this manuscript. That is, whenever R is a noetherian commutative ring there is an equivalence $\underline{\mathrm{D}}^{\mathrm{b}}(A) \simeq \mathrm{D_{sg}}(\operatorname{Spec} A)$. This lead to a significant, renewed interest in this work of Buchweitz in the years after [**107**] appeared.

A startling, very useful result on singularity categories is Orlov's theorem that closely relates the bounded derived category of sheaves on a Gorenstein projective variety with the corresponding graded singularity category.

To state this result we begin with a positively graded commutative algebra A over a field $A_0 = k$, such that A is noetherian and Gorenstein. The last hypothesis means that the bigraded vector space $\operatorname{Ext}_A(k, A)$ is one dimensional, concentrated in cohomological degree $d = \dim A$ and in some internal degree a, known as the *a-invariant* of A. In the important case $a = 0$, A is known as a *Calabi–Yau algebra.*

The theorem also involves a choice of integer i, with which one defines a diagram of adjoint functors

$$\underline{\mathrm{D}}^{\mathrm{b}}_{\mathrm{gr}}(A) \underset{\pi}{\overset{b_i}{\rightleftarrows}} \mathrm{D}^{\mathrm{b}}_{\mathrm{gr}_{\geqslant i}}(A) \underset{\mathrm{R}\Gamma_{\geqslant i}}{\overset{\widetilde{(-)}}{\rightleftarrows}} \mathrm{D}^{\mathrm{b}}(\operatorname{Proj}\text{-}A)\,.$$

Here $\mathrm{D}^{\mathrm{b}}_{\mathrm{gr}_{\geqslant i}}(A)$ is the bounded derived category of graded A-modules concentrated in internal degrees i and above; see [**109**] and also [**41**] for details on these functors. For the notion of a semi-orthogonal decomposition see [**69**, Section 1.4].

THEOREM C.3.1 ([**109**, Theorem 2.13]). *In the situation described above*

(a) *if $a > 0$ then $\widetilde{b_i}\colon \underline{\mathrm{D}}^{\mathrm{b}}_{\mathrm{gr}}(A) \to \mathrm{D}^{\mathrm{b}}(\operatorname{Proj}\text{-}A)$ is fully faithful and there is a semi-orthogonal decomposition*

$$\mathrm{D}^{\mathrm{b}}(\operatorname{Proj}\text{-}A) = \big(\mathcal{O}(-i-a+1), \dots, \mathcal{O}(-i), \widetilde{b_i}(\underline{\mathrm{D}}^{\mathrm{b}}_{\mathrm{gr}}(A))\big)\,;$$

(b) *if $a = 0$ the following functors are inverse equivalences*

$$\underline{\mathrm{D}}^{\mathrm{b}}_{\mathrm{gr}}(A) \underset{\pi\mathrm{R}\Gamma_{\geqslant i}}{\overset{\widetilde{b_i}}{\rightleftarrows}} \mathrm{D}^{\mathrm{b}}(\operatorname{Proj}\text{-}A)\,;$$

(c) *if $a < 0$ then $\pi\mathrm{R}\Gamma_{\leqslant i}\colon \mathrm{D}^{\mathrm{b}}(\operatorname{Proj}\text{-}A) \to \underline{\mathrm{D}}^{\mathrm{b}}_{\mathrm{gr}}(A)$ is fully faithful and there is a semi-orthogonal decomposition*

$$\underline{\mathrm{D}}^{\mathrm{b}}_{\mathrm{gr}}(A) = \big(k(-i-a), \dots, k(-i+1), \pi\mathrm{R}\Gamma_{\leqslant i}(\mathrm{D}^{\mathrm{b}}(\operatorname{Proj}\text{-}A))\big)\,.$$

To give a simple example, if one takes A to be a standard graded polynomial algebra $k[x_0, \dots, x_n]$, then the a-invariant is equal to $n+1$ and $\underline{\mathrm{D}}^{\mathrm{b}}_{\mathrm{gr}}(A) \simeq 0$. In this case the theorem asserts that $\mathrm{D}^{\mathrm{b}}(\operatorname{Coh}(\mathbb{P}^n_k)) = \mathrm{D}^{\mathrm{b}}(\operatorname{Proj}\text{-}A) = \big(\mathcal{O}(-n), \dots, \mathcal{O}(0)\big)$ is generated by a full, strong, exceptional collection, recovering a famous result of Beĭlinson [**25**].

Orlov's work itself is based on ideas of *homological mirror symmetry*, and this is explored further in physics by, for example, Herbst, Hori and Page in [**66**]. The theorem has also inspired much work in algebraic geometry, [**19, 20, 22, 64, 85**] commutative algebra [**42**], and representation theory [**94**], to name a few. Pavlov explains in [**111**] how Orlov's equivalence can be exploited to compute the Betti numbers of projectively embedded Calabi–Yau varieties, and in particular he obtains explicit formulas for the Betti numbers of an elliptic curve embedded in $\mathbb{P}^2_k$, when k is an algebraically closed field.

The arguments of Orlov have been clarified and extended to the non-commutative context by Burke and Stevenson in [**41**], opening up applications to non-commutative projective geometry. These authors acknowledge as their inspiration a sequence of lectures given by Buchweitz in Bielefeld in 2008; see also the lectures he gave at the Banff International Research Station [**35, 36**].

C.4 Duality. In Chapter 7 of this manuscript, Buchweitz explores various types of duality phenomena over Gorenstein rings. In this part we discuss some of the mathematics that grew out of it, acknowledging at the same time that there are yet many ideas in Chapter 7, notably those concerning maps of rings, that remain to be explored fully. We begin by recalling the language of Serre functors which helps to bring out the commonality between the duality statements to be encountered below.

(C.4.1) *Serre functors.* Let S be a commutative noetherian semi-local ring, with radical $\mathfrak{n}$, and let E be an injective hull of $S/\mathfrak{n}$. We also fix an S-linear category C; in what follows this will usually be triangulated. Roughly speaking, the linearity condition means that for each pair of objects X, Y in C there is a natural S-action on $\operatorname{Hom}_{\mathsf{C}}(X,Y)$, compatible with compositions in C. We say C is *Hom-finite* if the length of the S-module $\operatorname{Hom}_{\mathsf{C}}(X,Y)$ is finite for all X, Y. Following Bondal and Kapranov [**31**] a *Serre functor* on such a category C is an auto-equivalence $\mathbb{S}\colon \mathsf{C} \to \mathsf{C}$ along with natural isomorphisms

$$\operatorname{Hom}_S(\operatorname{Hom}_{\mathsf{C}}(X,Y),E) \cong \operatorname{Hom}_{\mathsf{C}}(Y,\mathbb{S}X)$$

that satisfy various compatibility conditions; see [**31**], [**69**, Definition 1.28], or [**114**], for details. It is helpful to repackage the data of a Serre functor into an S-linear trace map

$$\tau_X\colon \operatorname{Hom}_{\mathsf{C}}(X,\mathbb{S}X) \longrightarrow E\,.$$

Composition in C followed by this trace map induces a bilinear pairing

$$\operatorname{Hom}_{\mathsf{C}}(Y,\mathbb{S}X) \otimes_S \operatorname{Hom}_{\mathsf{C}}(X,Y) \longrightarrow E\,;$$

this pairing has to be non-degenerate.

The prototypical example is the category of coherent sheaves on a projective space over a field k, and then $\mathbb{S}$ is given by tensoring with a shift of the dualizing sheaf. In this case $S = k$, and the trace map is the classical residue map; see, for example, [**92**].

(C.4.2) *Isolated singularities.* Let S be a commutative Gorenstein local ring with an isolated singularity. This last condition means that the global dimension of the ring $S_{\mathfrak{p}}$ is finite for all non-maximal primes $\mathfrak{p}$ in $\operatorname{Spec} S$; equivalently, for all MCM S-modules M, N and integers i, the Tate cohomology modules $\underline{\operatorname{Ext}}^i_S(M,N)$ have finite length as S-modules; see Lemma 7.7.4.

Recast in the language introduced above, a theorem of Auslander [**3**] states that the $(1-d)$th stable syzygy functor Ω^{1-d}, where d is the dimension of S, is a Serre functor on $\underline{\mathrm{MCM}}(S)$, the stable category of MCM S-modules. In Chapter 7,

Buchweitz establishes a number of statements that generalise this "Auslander duality"; see Theorem 7.7.5 for one version. In the remark following Theorem 7.7.5 he asks for an explicit description of the trace map in this context, which in this case takes the form

$$\tau_N\colon \underline{\mathrm{Hom}}_S(N, \Omega^{1-d}N) \longrightarrow E\,.$$

Buchweitz suggests that such a description is likely to involve Grothendieck's theory of residues and differential forms. Subsequently Murfet [**102**] gave just such a description of the trace map. It has a particularly simple description for hypersurface rings; we sketch this below. It involves the theory of local cohomology and the associated calculus of generalized fractions; see [**92**].

Let k be a field of characteristic zero and $P := k[\![x_1, \dots, x_n]\!]$ the ring of formal power series over k in indeterminates $x_1, \dots, x_n$. Then P is a commutative noetherian local ring with maximal ideal $\mathfrak{n} := (x_1, \dots, x_n)$ and residue field k. The top local cohomology module, $\mathrm{H}^n_{\mathfrak{n}}(P)$, of P with support in $\mathfrak{n}$ is an injective hull of k as a P-module. This comes equipped with a k-linear *residue map*

$$\mathrm{Res}_{P/k}\colon \mathrm{H}^n_{\mathfrak{n}}(P) \longrightarrow k\,.$$

As a k-vector space the local cohomology module has a basis consisting of "generalized" fractions, also called inverse polynomials:

$$\left[\frac{1}{x_1^{a_1}\cdots x_n^{a_n}}\right] \quad \text{where } a_i \geq 1 \text{ for each } i.$$

The P-action defined by

$$x_i \cdot \left[\frac{1}{x_1^{a_1}\cdots x_n^{a_n}}\right] = \left[\frac{1}{x_1^{a_1}\cdots x_i^{a_i-1}\cdots x_n^{a_n}}\right]$$

where it is understood that the fraction on the right is zero if $a_i = 1$. In particular, the socle of $\mathrm{H}^n_{\mathfrak{n}}(P)$ is spanned by the element

$$\left[\frac{1}{x_1\cdots x_n}\right].$$

The residue map on $\mathrm{H}^n_{\mathfrak{n}}(P)$ is determined by

$$\mathrm{Res}_{P/k}\left[\frac{1}{x_1^{a_1}\cdots x_n^{a_n}}\right] = \begin{cases} 1 & \text{when } a_i = 1 \text{ for each } i \\ 0 & \text{otherwise.}\end{cases}$$

This construction depends on the choice of power series generators for P. For applications, in particular to get a functorial definition of the residue map, one has to work with the local cohomology of a suitable module of differential forms. What is more, it is possible, and important, to define generalized fractions and the residue map with respect to any system of parameters $t_1, \dots, t_n$ for the ring P; see [**92**, Chapter 5] for details.

Suppose $S := P/I$, for some ideal $I \subseteq \mathfrak{n}$. Then if E is an injective hull of k as a P-module, the S-module $\mathrm{Hom}_P(S, E)$ is an injective hull of k as an S-module. This module consists of elements of E annihilated by I, namely

$$E_S := \mathrm{Hom}_P(S, E) = \{e \in E \mid I \cdot e = 0\}\,.$$

In particular, setting $E = \mathrm{H}^n_{\mathfrak{n}}(P)$, and restricting the residue map defined above to E_S yields the residue map

$$\mathrm{Res}_{S/k}\colon E_S \longrightarrow k\,.$$

Let now $S := P/(W)$, where W has an isolated singularity at the origin. The Jacobian criterion permits one to express this last condition in terms of the partial derivates $\partial_i W$ of W with respect to x_i. Namely, W has an isolated singularity at the origin precisely when the radical of the ideal $J(W)$ generated by the partial derivatives $\partial W := \partial_1 W, \ldots, \partial_n W$ equals $\mathfrak{n}$; in other words, ∂W is a system of parameters for P. This condition is also equivalent to the Jacobian algebra $P/J(W)$ being finite dimensional over k. The upshot is that one can express the local cohomology module $\mathrm{H}^n_{\mathfrak{n}}(P)$ in terms of generalized fractions in ∂W.

A *matrix factorization* of W is a pair of square matrices A, B, say of size s, over P with $AB = WI_s = BA$. Given such a matrix factorization $\mathrm{Cok}(A)$ is a maximal Cohen–Macaulay S-module, and every MCM S-module occurs thus; see [**Eis80**], and also [**128**]. For the present purpose, it is more convenient to view a matrix factorization of W as a degree one map $d_X \colon X \to X$ of the $\mathbb{Z}/2\mathbb{Z}$-graded finite free P-module $X = X_0 \oplus X_1$, satisfying $d_X^2 = W\,\mathrm{id}$. Under this identification a matrix factorization (A, B) of W corresponds to the map

$$\begin{bmatrix} 0 & A \\ B & 0 \end{bmatrix} : P^s \oplus P^s \longrightarrow P^s \oplus P^s$$

of the $\mathbb{Z}/2\mathbb{Z}$-graded free P-module consisting of P^s in degrees 0 and 1. There is an obvious notion of a morphism of matrix factorizations, and homotopies between them. The corresponding homotopy category is equivalent to the stable category of MCM S-modules. Under this equivalence, the syzygy functor Ω on $\underline{\mathrm{MCM}}(S)$ corresponds to the map that assigns to $X := (X_0 \oplus X_1, d_X)$ the object $\Omega X := (X_1 \oplus X_0, -d_X)$. Thus if N is the MCM module corresponding to X, a class in $\underline{\mathrm{Hom}}_S(N, \Omega^{2-n} N)$ is represented by a morphism $\psi \colon X \to \Omega^{2-n} X$ of matrix factorizations. Recall that the dimension of S is $n-1$ and $\Omega^2 = \mathrm{id}$.

With these identifications Murfet [**102**] proves that the trace map

$$\tau_N \colon \underline{\mathrm{Hom}}_S(N, \Omega^{2-n} N) \longrightarrow k$$

is given by the assignment

$$\tau_N(\psi) = \frac{(-1)^{\binom{n-1}{2}}}{n!} \mathrm{Res}_{S/k} \begin{bmatrix} \mathrm{str}_S(\psi \circ d_{S/k}(dX)^n) \\ \partial_1 W, \ldots, \partial_n W \end{bmatrix} .$$

Here $d_{S/k}(dX)^n$ is a signed symmetric sum of the derivatives of the map d_X with respect to the x_i, and str is the "super-trace" of the $\mathbb{Z}/2\mathbb{Z}$-graded map $X \to X$ obtained by composing $d_{S/k}(dX)^n$ with ψ. The terms in the formula require further explanation; see [**102**, §6] for details, and also [**53**].

(C.4.3) *Gorenstein algebras.* We move on to not-necessarily commutative algebras. To begin with, let A be a Gorenstein Artin algebra. The Artin condition means that A is finite over its center, say S, which is semi-local and artinian, and so $\mathrm{mod}(A)$ is an S-linear Hom-finite category. The duality theorem of Auslander [**3**] is equivalent to the statement that the stable category $\underline{\mathrm{MCM}}(A)$ of MCM A-modules has a Serre functor, namely:

$$\mathbb{S}(M) := \Omega^{-1}(D\mathrm{Tr}M)) \quad \text{for } M \in \mathrm{MCM}(A)$$

where Tr is the Auslander transpose 6.4 and $D := \mathrm{Hom}_S(-, E)$, with E as before; see (C.4.1). Buchweitz's Theorem 7.7.5 generalises this to a larger family of Gorenstein algebras with isolated singularities.

The duality theorem of Auslander is central to the representation theory of finite dimensional algebras. For example, it implies the existence of almost-split sequences, in the sense of Auslander and Reiten [**9**]. In fact, for any Hom-finite S-linear category with the Krull–Remak–Schmidt property, the existence of a Serre functor implies the existence of AR sequences in that category. The relationship between these notions has been clarified by Reiten and Van den Bergh [**114**]. This is one reason for the interest in establishing duality results akin to those of Auslander and Buchweitz.

Auslander [**3**] proved a "defect formula" for general associative rings from which one can deduce the duality theorems, at least for finite dimensional algebras; see also [**8**]. In [**87**] Krause gives a simple proof of this defect formula and in [**89**] he extends this to the context of general Grothendieck categories.

In a different direction, there is recent work of Iyengar and Krause [**73**] where one can find duality theorems that apply to any Gorenstein algebra A that is finite and projective over its center. The point is that there is no a priori assumption that A has isolated singularities.

Returning to Buchweitz, the results in Chapter 7 are stated and proved in the relative case, that is to say, they concern *maps* of rings. This is in keeping with the spirit of developments in commutative algebra and algebraic geometry initiated by Grothendieck. As far as we know, these ideas of Buchweitz have not been developed further; see however the works of Efimov and Positselski [**54**], Burban and Kalck [**40**], and Kalck and Yang [**80**], that deal with relative singularity categories.

(C.4.4) *Gorenstein categories.* In Section 8.2 Buchweitz raises the prospect of a general theory of Gorenstein categories encompassing the results presented in the manuscript. One candidate for such a theory is introduced by Benson, Iyengar, Krause and Pevtsova [**27**]. The framework is distinctly inspired by and geared towards applications to the representation theory of finite groups and finite group schemes. For applications to algebraic topology there is work of Dwyer, Greenlees, and Iyengar [**52**] but this is closer to the commutative theory of Gorenstein rings, and in particular to local cohomology theorems in that context. These ideas are closely connected with Grothendieck duality theory in algebraic geometry. Some of those results have been extended to tensor triangulated categories; see, for example, Balmer, Dell'Ambrogio and Sanders [**21**].

C.5 Beyond Gorenstein rings. After defining MCM modules over Iwanaga–Gorenstein rings, that is to say, strongly Gorenstein rings, Buchweitz notes that in this case they are the G-projective modules, defined by Auslander and Bridger [**AB69**] for a larger class of rings. He goes on to state that "in dealing only with strongly Gorenstein rings, we believe 'MCM' to be more suggestive." In the influential paper [**5**], Auslander and Buchweitz pursued that path and developed a categorical context based on the class of maximal Cohen–Macaulay modules over commutative local Cohen–Macaulay rings.

In this section we sketch some developments along the road not taken.

(C.5.1) *Gorenstein projectives.* Let S be a two-sided noetherian ring, $\mathrm{P}(S)$ the full subcategory of projective modules in $\mathrm{mod}(S)$, and $\mathrm{GP}(S)$ that of the G-projective modules introduced in Definition B.1.2: A finitely generated S-module M is G-projective if

(1) $\mathrm{Ext}^i_S(M, S) = 0$ for $i \geq 1$,
(2) $\mathrm{Ext}^i_{S^{\mathrm{op}}}(M^*, S) = 0$ for $i \geq 1$, and
(3) $M \to M^{**}$ is bijective.

Evidently $\mathrm{GP}(S) \supseteq \mathrm{P}(S)$. Determining if a given module M is G-projective may be complicated, even when every projective in $\mathrm{mod}(S)$ is free. This is due to conditions (1) and (2) in the definition. Avramov and Martsinkovsky [**17**] showed that if these two are satisfied, and S is a commutative graded connected algebra over a field k, then condition (3) can be weakened to requiring that the biduality map be injective or surjective. They asked if the definition of G-dimension contains redundancies. When k is not algebraic over a finite field, Jorgensen and Şega [**79**] exhibited an algebra S as above, with reflexive modules M and N satisfying $\mathrm{Ext}^i_S(M, R) = 0 \neq \mathrm{Ext}^i_S(M^*, R)$ and $\mathrm{Ext}^i_S(N^*, R) = 0 \neq \mathrm{Ext}^i_S(N, R)$ for $i \geq 1$; the ring in the example satisfies $\mathrm{rank}_k(S) = 8$ and is Koszul. It still is not known if (1) and (2) imply (3).

A central open problem concerning G-projectives is to relate the condition $\mathrm{GP}(S) = \mathrm{P}(S)$ to the structures of S and $\mathrm{mod}(S)$. Some results are available for commutative rings S that are local or graded connected. If $S = Q/(f)$ for some ring Q of the same type as S and f a non-zero-divisor contained in the square of the maximal ideal of Q, respectively, irrelevant maximal ideal, then matrix factorizations of f provide G-projective modules that are not projective; see Avramov, Gasharov, and Peeva [**12**]. On the other hand, if S is Golod and not a hypersurface, then $\mathrm{GP}(S) = \mathrm{P}(S)$ holds; see Avramov and Martsinkovsky [**17**].

Only two Cohen–Macaulay local rings of codimension bigger than 2 and of finite Cohen–Macaulay type (that is to say, it has only finitely many isoclasses of indecomposable maximal Cohen–Macaulay modules) are known; they were discovered by Auslander and Reiten [**7**] and have minimal multiplicity. In particular, these rings are Golod and hence satisfy $\mathrm{GP}(S) = \mathrm{P}(S)$. They are completions of graded connected k-algebras R of finite graded Cohen–Macaulay type with $\dim R = 3$; see Auslander and Reiten [**6**] and Solberg [**119**]. Eisenbud and Herzog [**55**] proved that when k is algebraically closed no other such algebras exist and conjectured that all local rings of finite Cohen–Macaulay type have minimal multiplicity; see also [**128**, Ch. 17] or [**95**, Ch. 16].

The preceding discussion reveals how $\mathrm{GP}(S)$ and $\mathrm{MCM}(S)$ diverge over non-Gorenstein rings. The size of the gap is underscored in a striking dichotomy, conjectured by Takahashi in [**122**], proved by him with separate arguments in dimensions 0, 1, and 2, and in general by Christensen, Piepmeyer, Striuli, and Takahashi [**46**]:

THEOREM C.5.2. *If a commutative local ring S has finitely many isoclasses of indecomposable G-projective modules, then S is Gorenstein or* $\mathrm{GP}(S) = \mathrm{P}(S)$.

In view of the results surveyed in C.2, when the residue field of S is algebraically closed and S has more than one, but finitely many isoclasses of indecomposable G-projective modules, the theorem implies then $\widehat{S}$ is an isolated hypersurface singularity, and a simple singularity if S is complete; see also [**95**, Ch. 12].

(C.5.3) *Finite G-dimension.* For S as in (C.5.1), Auslander and Bridger [**AB69**] defined the *G-dimension* of $M \in \mathrm{mod}(S)$, denoted by $\text{G-dim}(M)$, as the infimum of the lengths of the resolutions of M by G-projectives. This invariant refines projective dimension: $\text{G-dim}_S(M) \leq \mathrm{projdim}_S(M)$ holds with equality if $\mathrm{projdim}_S(M)$ is finite. Thus $\text{G-dim}_S(M) = 0$ means that M is G-projective, and $\text{G-dim}_S(M)$

can be computed from any projective resolution of M. If S is Iwanaga–Gorenstein, $\text{G-dim}_S(M) \le \text{vdim}\, S$ holds for all $M \in \text{mod}(S)$; see Lemma 4.2.2(iv).

We write G-mod(S) for the full subcategory of mod(S) consisting of the modules of finite G-dimension. A sequence E of homomorphisms in mod(S) is said to be *proper* if the sequence $\text{Hom}_S(L, E)$ is exact for every $L \in \text{GP}(S)$; taking $L = S$ shows that proper sequences are exact. A resolution P of M is said to be proper if the mapping cone of the augmentation $P \to M$ is a proper sequence.

A number of *properties* of modules over Iwanaga–Gorenstein rings translate into *characterizations* of modules of finite G-dimension, once the relevant notions for an S-module M have been recalibrated; we do that following [**17**]. A *G-approximation* of M is an exact sequence

$$0 \longrightarrow M' \longrightarrow L \longrightarrow M \longrightarrow 0$$

with $L \in \text{GP}(S)$ and $\text{projdim}_S(M') < \infty$. A complex C is said to be *totally acyclic* if it consists of finitely generated projectives and satisfies the conditions: $\text{H}_i(C) = 0$ and $\text{H}_i(\text{Hom}_S(C, S)) = 0$ for each i. A *complete resolution* of M is a morphism on complexes $\mathbf{CR}(M) \to \mathbf{P}(M)$ such that $\mathbf{CR}(M)$ is totally acyclic, $\mathbf{P}(M) \simeq M$ is a projective resolution, and $\mathbf{CR}(M)_i \to \mathbf{P}(M)_i$ is bijective for $i \gg 0$.

The condition $\text{G-dim}_S(M) \le g$ is equivalent to each one of the conditions:

(i) M has G-approximation M' with $\text{projdim}_S(M') \le g - 1$.
(ii) M has a proper resolution $\mathbf{G}(M)$ of length g with $\mathbf{G}(M)_i \in \text{GP}(S)$ for all i.
(iii) M has a complete resolution $\mathbf{CR}(M) \to \mathbf{P}(M)$ with $\mathbf{CR}(M)_i \xrightarrow{\cong} \mathbf{P}(M)_i$ for $i \ge g$.

The equivalences are proved by producing one type of resolution from another by means of explicit constructions. Various arrangements of similar arguments have been used at least since Tate's introduction of his cohomology theory for finite groups, in [**AB69**], and in Chapters 4 and 5 of this book when S is Iwanaga–Gorenstein and M is MCM; compare part (ii) in the proof of Theorem 4.4.1 to (iii) and the proof of Theorem 5.1.21 to (i).

After fixing appropriate resolutions for each M, for all $N \in \text{mod}(S)$ and $i \in \mathbb{Z}$ define the ith *relative* Ext and the ith *Tate* Ext of M and N over S, by setting

$$\text{Ext}^i_{\text{GP}(S)}(M, N) := \text{H}^i\left(\text{Hom}_S(\mathbf{G}(M), N)\right) \quad \text{and}$$
$$\widehat{\text{Ext}}^i_S(M, N) := \text{H}^i\left(\text{Hom}_S(\mathbf{CR}(M), N)\right),$$

respectively. By choosing a comparison of resolutions $\mathbf{P}(M) \to \mathbf{G}(M)$ and invoking the morphism $\mathbf{CR}(M) \to \mathbf{P}(M)$ embedded in the definition, one obtains maps

$$\text{Ext}^i_{\text{GP}(S)}(M, N) \longrightarrow \text{Ext}^i_S(M, N) \longrightarrow \widehat{\text{Ext}}^i_S(M, N) \quad \text{for all} \quad i \in \mathbb{Z}\,.$$

The defining properties of the resolutions are utilized for obtaining comparison morphisms of complexes that are unique up to homotopy. As a consequence, the abelian groups and maps above do not depend on the choices in their constructions and generally behave like values of cohomological functors on $\text{G-mod}(S) \times \text{mod}(S)$, even though G-mod($S$) is not abelian; see [**17**]. The main result of that paper is that they yield an exact sequence, natural in both module arguments:

$$\begin{aligned} 0 &\longrightarrow \text{Ext}^1_{\text{GP}(S)}(M, N) \longrightarrow \text{Ext}^1_S(M, N) \longrightarrow \widehat{\text{Ext}}^1_S(M, N) \\ &\longrightarrow \text{Ext}^2_{\text{GP}(S)}(M, N) \longrightarrow \quad \cdots \quad \longrightarrow \widehat{\text{Ext}}^{g-1}_S(M, N) \\ &\longrightarrow \text{Ext}^g_{\text{GP}(S)}(M, N) \longrightarrow \text{Ext}^g_S(M, N) \longrightarrow \widehat{\text{Ext}}^g_S(M, N) \longrightarrow 0 \end{aligned}$$

and for all $i \geq g+1$ isomorphisms

$$\operatorname{Ext}_S^i(M,N) \xrightarrow{\cong} \widehat{\operatorname{Ext}}_S^i(M,N).$$

Explicit constructions are once again brought into play, this time in order to work around the non-abelian nature of G-mod(S) when constructing connecting maps.

The definition of G-dimension for modules, the construction of the groups $\widehat{\operatorname{Ext}}_S^i(M,N)$, and the proofs of their properties were extended by Veliche [**123**] to complexes M and N with finitely generated homology. In the special case when S is Iwanaga–Gorenstein, $\widehat{\operatorname{Ext}}_S^i(M,N)$ is equal to the Tate cohomology group $\underline{\operatorname{Ext}}_S^i(M,N)$ in Definition 6.1.1, by Lemma 6.1.2(ii). Also, the exact sequence above specializes to that in Theorem 6.2.5(3); this is explained in the next subsection.

(C.5.4) *Stable cohomology.* In the mid 1980s, Pierre Vogel constructed a $\mathbb{Z}$-graded cohomology theory that is defined for all modules over an arbitrary associative ring S and extends Tate cohomology for finite groups. His theory, also called Tate–Vogel cohomology, has been described and applied by Goichot [**60**], Martsinkovsky [**97**], and Avramov and Veliche [**18**], among others. Constructions of equivalent theories are mentioned at the end of this subsection.

We present Vogel's construction and some of its properties following [**18**, Sections 1–4], with modules replaced by bounded complexes, at no cost to the reader. For any two complexes F, G define a set of *bounded* homomorphisms by setting

$$\overline{\operatorname{Hom}}_S(F,G) := \{\alpha \in \operatorname{Hom}_S(F,G) \mid \alpha(F_i) = 0 \text{ for all } |i| \gg 0\}.$$

It is a subcomplex of $\operatorname{Hom}_S(F,G)$, with component of degree n equal to

$$\overline{\operatorname{Hom}}_S(F,G)^n = \coprod_{i\in\mathbb{Z}} \operatorname{Hom}_S(F^i, G^{i+n}) \subseteq \prod_{i\in\mathbb{Z}} \operatorname{Hom}_S(F^i, G^{i+n}) = \operatorname{Hom}_S(F,G)^n.$$

Given bounded complexes M, N, choose resolutions $\mathbf{P}(M)$ and $\mathbf{P}(N)$ lying in $\mathrm{K}^{-,\mathrm{b}}(\mathrm{P}(S))$, and form the exact sequence of complexes of abelian groups

(C.5.5)

$$0 \to \overline{\operatorname{Hom}}_S(\mathbf{P}(M),\mathbf{P}(N)) \to \operatorname{Hom}_S(\mathbf{P}(M),\mathbf{P}(N)) \to \widehat{\operatorname{Hom}}_S(\mathbf{P}(M),\mathbf{P}(N)) \to 0$$

with the inclusion map on the left and the quotient complex on the right. The group

$$\widehat{\operatorname{Ext}}_S^i(M,N) := \mathrm{H}^i\left(\widehat{\operatorname{Hom}}_S(\mathbf{P}(M),\mathbf{P}(N))\right)$$

is the ith *stable cohomology* group of (M,N) (called *Tate–Vogel cohomology* in [**60**]).

If S is two-sided noetherian and M has finite G-dimension, then the groups $\widehat{\operatorname{Ext}}_S^i(M,N)$ defined in this subsection and the preceding one are canonically isomorphic. Indeed, the argument devised by Vogel for the Tate cohomology of finite groups works in general, and is reproduced in [**18, 60, 96**].

As in [**18**], we define the *bounded cohomology* of (M,N) in degree i to be

$$\overline{\operatorname{Ext}}_S^i(M,N) := \mathrm{H}^i(\overline{\operatorname{Hom}}_S(\mathbf{P}(M),\mathbf{P}(N))).$$

The exact sequence of complexes yields an exact sequence of graded abelian groups

$$\text{(C.5.6)}\qquad \begin{aligned} \cdots \to \overline{\operatorname{Ext}}_S(M,N) \to \operatorname{Ext}_S(M,N) \to \widehat{\operatorname{Ext}}_S(M,N)\\ \to \overline{\operatorname{Ext}}_S(M,N)[1] \to \cdots. \end{aligned}$$

One gets more out of this exact sequence when S is noetherian and $\mathrm{H}(M)$ is finitely generated. Since $\mathbf{P}(M)$ lies in $\mathrm{K}^{-,\mathrm{b}}(\mathrm{P}(S))$, the natural map

$$\mathbf{P}(M)^* \otimes_S \mathbf{P}(N) \longrightarrow \mathrm{Hom}_S(\mathbf{P}(M), \mathbf{P}(N))$$

factors through an isomorphism $\mathbf{P}(M)^* \otimes_S \mathbf{P}(N) \xrightarrow{\cong} \overline{\mathrm{Hom}}_S(\mathbf{P}(M), \mathbf{P}(N))$, yielding

$$\mathrm{Tor}^S_{-i}(\mathrm{RHom}_S(M,S), N) \xrightarrow{\cong} \overline{\mathrm{Ext}}^i_S(M,N)\,. \tag{C.5.7}$$

Here we view $\mathrm{RHom}_S(M,S)$ as an object of $\mathrm{D}(S^{\mathrm{op}})$ in the usual way. Its cohomology in degree i is $\mathrm{Ext}^i_S(M,S)$, so $\mathrm{RHom}_S(M,S)$ is in $\mathrm{D}^{\mathrm{b}}(S^{\mathrm{op}})$ exactly when $\mathrm{Ext}^i_S(M,S) = 0$ for $i \gg 0$. This is satisfied when S is Iwanaga–Gorenstein, and then the exact sequence (C.5.6) specializes to that from Theorem 6.2.5(3).

Multiplicative structures are ubiquitous in cohomology. In Chapter 7 Buchweitz discusses bimodule structures in the context of Tate cohomology over Iwanaga–Gorenstein rings. By inception, stable cohomology is ideally suited for defining various products and working with them. Some of them are described below, and a sample application is presented in ((C.5.9)).

For starters, given complexes L, M, N of S-modules, the *composition pairing*

$$\mathrm{Hom}_S(\mathbf{P}(M), \mathbf{P}(N)) \times \mathrm{Hom}_S(\mathbf{P}(L), \mathbf{P}(M)) \longrightarrow \mathrm{Hom}_S(\mathbf{P}(L), \mathbf{P}(N))$$

defines dg rings $\mathbf{E}(M) := \mathrm{Hom}_S(\mathbf{P}(M), \mathbf{P}(M))$ and $\mathbf{E}(N) := \mathrm{Hom}_S(\mathbf{P}(N), \mathbf{P}(N))$ with a right, respectively, left actions on $\mathrm{Hom}_S(\mathbf{P}(M), \mathbf{P}(N))$ and turn it into a dg bimodule. If $f\colon M \to N$ or $g\colon L \to M$ is bounded, then so is $f \circ g$, and hence $\overline{\mathrm{Hom}}_S(\mathbf{P}(M), \mathbf{P}(N))$ is a dg sub-bimodule. Thus (C.5.5) is an exact sequence of dg bimodules. In homology one gets graded rings $\mathrm{Ext}_S(M,M) := \mathrm{H}(\mathbf{E}(M))$ and $\mathrm{Ext}_S(N,N) := \mathrm{H}(\mathbf{E}(N))$ and an exact sequence (C.5.6) of bimodules on which they act from the right and from the left, respectively. As $\overline{\mathrm{Hom}}_S(\mathbf{P}(M), \mathbf{P}(M))$ is a two-sided dg ideal of $\mathbf{E}(M)$, the map

$$\mathbf{E}(M) \longrightarrow \overline{\mathrm{Hom}}_S(\mathbf{P}(M), \mathbf{P}(M))$$

is a homomorphism of dg rings and induces a map of graded rings:

$$\mathrm{Ext}_S(M,M) \longrightarrow \widehat{\mathrm{Ext}}_S(M,M)\,.$$

These interlocking multiplicative structures are invaluable in computations. A helpful general observation in [**18**] is that $\overline{\mathrm{Ext}}_S(M,N)$ is a torsion module for the ideal $\mathrm{Ext}^{\geqslant 1}_S(N,N)$: For every $\tau \in \overline{\mathrm{Ext}}_S(N,N)$ there exists an integer $j \geq 1$ such that $\mathrm{Ext}^{\geqslant j}_S(N,N)\tau = 0$; on the other hand, the action of $\mathrm{Ext}_S(M,M)$ may not have the corresponding property. Buchweitz discusses such bimodule structures over Iwanaga–Gorenstein rings in 7.1.

Other stable cohomology theories were proposed before accounts of Vogel's theory [**60**] for general rings and of Buchweiz's work for Iwanaga–Gorenstein rings were published. Benson and Carlson [**28**] gave two new constructions of Tate cohomology of finite groups; it was realized later that both of them can be used over any ring. A stable cohomology theory based on stabilization of satellites was introduced by Mislin [**101**], who also proved that his theory is equivalent to one of those in [**28**]; his starting point is a stabilization isomorphism, identical to that in Proposition 6.5.1. Kropholler [**90**] surveys the constructions in [**60**], [**28**], and [**101**] and points out that "the other" construction in [**28**] is equivalent to Vogel's. Nucinkis [**106**] used injective modules, resolutions, and satellites in lieu of projective ones to construct "dual" versions of all three theories, proved the equivalence of those versions, and determined the rings over which the injective and projective theories concur. Incidentally, dualizing the equivalence proofs in [**106**] provides an optimal path for proving the equivalence of the original theories.

The existence of different avatars of stable cohomology has facilitated its uses in various contexts. For applications in algebraic geometry see Orlov [**107**, **108**, **109**], and also C.3; recent applications in algebra include the papers of Emmanouil and Talelli [**56**, **57**] and of Mazza and Symonds [**99**]. One feature of Vogel's construction is that it can easily be extended to the context of differential graded modules over differential graded algebras. This can be useful even if one is only interested in rings and modules; see [**13**] for an application to computing Bass series of certain commutative local rings.

(C.5.8) *Stable categories.* Buchweitz focuses on Iwanaga–Gorenstein rings, but $\underline{\mathrm{D}}^{\mathrm{b}}(S) := \mathrm{D}^{\mathrm{b}}(S)/\,\mathrm{D}^{\mathrm{b}}_{\mathrm{perf}}(S)$ is a sensible concept of stable derived category for every (right) noetherian ring S. This is how it is introduced in Definition 1.2.2, except that Buchweitz assumes S is noetherian on both sides. Following Orlov [**107**] this has come to be known as the *singularity category* of S; see C.3.

One of the first things to understand about $\underline{\mathrm{D}}^{\mathrm{b}}(S)$ is the abelian group of morphisms from M to $N[i]$. When S is Iwanaga–Gorenstein, Buchweitz uses it to *define* Tate cohomology group $\widehat{\mathrm{Ext}}^{i}_{S}(M,N)$; cf. Definition 6.1.1. From that point of view, the techniques and results concerning Tate cohomology, to which Chapter 6 is devoted, become results about homomorphisms in $\underline{\mathrm{D}}^{\mathrm{b}}(S)$.

Over an arbitrary (right) noetherian ring S, stable cohomology takes over the role of Tate cohomology, as the preserved notation suggests. Indeed, Beligiannis [**26**, Definition 3.1 and Corollary 3.9(1)] proved that $\widehat{\mathrm{Ext}}^{0}_{S}(M,N)$ is the group of morphisms in a suitable stablization of the category of S-modules; the assertion for every i follows by degree shifting, and with these identifications the pairing of stable cohomology groups defined in ((C.5.4)) becomes composition of homomorphisms. When S is Iwanaga–Gorenstein this specializes to Theorem 6.2.5. In general, it shows that the following conditions are equivalent for any S-complex M with $\mathrm{H}(M)$ finitely generated:

(i) M is perfect, that is to say, has finite projective dimension;
(ii) $\mathrm{Hom}_{\underline{\mathrm{D}}^{\mathrm{b}}(S)}(M,N) = 0$ for every $N \in \underline{\mathrm{D}}^{\mathrm{b}}(S)$;
(iii) $\mathrm{Hom}_{\underline{\mathrm{D}}^{\mathrm{b}}(S)}(N,M) = 0$ for every $N \in \underline{\mathrm{D}}^{\mathrm{b}}(S)$;
(iv) $\mathrm{Hom}_{\underline{\mathrm{D}}^{\mathrm{b}}(S)}(M,M) = 0$.

Of course, these equivalencies can also be obtained directly from the construction of stable cohomology groups.

Computation of $\widehat{\mathrm{Ext}}^{i}_{S}(M,N)$ when $\text{G-dim}_S(M) = \infty = \mathrm{projdim}_S(N)$ tends to be complicated, not least because it is defined as a cohomology group of a complex of inherently enormous modules. As often in homological algebra, one looks for possibilities to stretch approaches that worked in some less threatening context. Here that would be modules M of finite G-dimension, whose stable cohomology groups can be computed from complete resolutions; in case S is a Noether algebra (see Definition B.1.5) they are finitely generated over its center, $\mathrm{Z}(S)$.

For any noetherian ring S, if $M \in \mathrm{mod}(S)$ has $\mathrm{Ext}^{\geqslant 1}_{S}(M,S) = 0$, then

$$\widehat{\mathrm{Ext}}^{i}_{S}(M,N) \cong \underline{\mathrm{Hom}}_{S}(\Omega^{i}M,N)$$

holds for every N in $\operatorname{mod}(S)$ and integer $i \geq 0$; see Lemma B.4.3, and also [**107**, Proposition 1.21]. As in $\underline{\mathrm{D}}^{\mathrm{b}}(S)$ one has $M \cong \Omega^s(M)[-s]$ for each $s \geq 0$, whenever $\operatorname{Ext}^i_S(M,S)=0$ holds for $i \gg 0$, the displayed isomorphism describes $\widehat{\operatorname{Ext}}^i_S(M,N)$ as a group of stable maps between certain finitely generated modules; again, that group is a finitely generated $\mathrm{Z}(S)$-module if S is a Noether algebra. See also [**18**, Proposition 3.1].

(C.5.9) *Commutative noetherian local rings.* Let S be a commutative noetherian local ring, k its residue field. The existence of minimal free resolutions for finitely generated S-modules allows one to strengthen the detection criteria for finite projective dimension recorded above as follows: For any S-complex M with $\mathrm{H}(M)$ finitely generated, if $\widehat{\operatorname{Ext}}^i_S(k,M)=0$ for *some* integer i, then M has finite projective dimension.

Set $g := \operatorname{depth} S$ and recall that $\operatorname{Ext}^i_S(k,S)=0$ for $i<g$. The ring S is regular if and only if $\operatorname{Ext}^i_S(k,k)=0$ for all, equivalently, some $i>g$. The ring S is Gorenstein if and only if $\operatorname{Ext}^i_S(k,S)\cong k$ and $\operatorname{Ext}^i_S(k,S)=0$ for $i>g$, if and only if $\operatorname{Ext}^i_S(k,S)=0$ for some $i>g$.

In the same vein, numerical invariants derived from stable cohomology can also be used to characterize singularity types. Switching to and stretching the notation in 10.1, we put

$$\mathrm{H}_S := \operatorname{Ext}_S(k,k) \quad\text{and}\quad \widehat{\mathrm{H}}_S := \widehat{\operatorname{Ext}}_S(k,k)\,.$$

The following results are proved in [**18**, Section 6] by using these graded k-algebras:

(1) S is regular if and only if $\widehat{\mathrm{H}}{}^n_S=0$ for all, equivalently, some n;
(2) S is Gorenstein if and only if $\operatorname{rank}_k \widehat{\mathrm{H}}{}^n_S<\infty$ for all, equivalently, some n;
(3) For $d:=\dim S$, $c:=\operatorname{codim} S$, and all n there is an inequality

$$\operatorname{rank}_k \widehat{\mathrm{H}}{}^n_S \geq \sum_{i=0}^{d}\binom{d}{i}\left(\binom{c+n-i-1}{c-1}+\binom{c+d-n-i-2}{c-1}\right),$$

with equality when S is complete intersection; equality for some $n\notin[d-2,1]$ implies that S is complete intersection.

The finiteness of $\operatorname{projdim}_S(k)$ characterizes regular rings, so the criteria for finiteness of projective dimension discussed above yield (1). The finiteness of each $\widehat{\mathrm{H}}{}^n_S$ when S is Gorenstein also comes from general properties of stable cohomology; see ((C.5.8)). The other implications rely on the exact sequence in the next result.

THEOREM C.5.10 ([**18**, Theorem 7.2]). *When S is not regular, there is an exact sequence of left graded modules over H_S, with ι from* (C.5.6), $\mathrm{H}^S := \operatorname{Hom}_k(\mathrm{H}_S,k)$ *with canonical action and grading, and H_S acting on the tensor product through H^S:*

$$0 \longrightarrow \mathrm{H}_S \xrightarrow{\ \iota\ } \widehat{\mathrm{H}}_S \longrightarrow \operatorname{Ext}_S(k,S)[1]\otimes_k \mathrm{H}^S \longrightarrow 0\,.$$

The exact sequence yields the following isomorphism in every degree n:

$$\operatorname{Cok}(\iota)^n \cong \bigoplus_{j\geq 0} \operatorname{Ext}^{n-g+1+j}_S(k,S)\otimes_k \operatorname{Hom}_k(\operatorname{Ext}^j_S(k,k),k)\,.$$

If S is not Gorenstein, then every direct summand is a tensor product of non-zero vector spaces, and hence has infinite k-rank; this finishes the proof of (2). If $\widehat{\mathrm{H}}{}^n_S=0$ for some n and S is not regular, then it has to be S is Gorenstein, by (2). Lower bounds on the ranks of H_S^n are known, and they are reached when S is complete intersection. Due to the exact sequence, these bounds yield ones for $\widehat{\mathrm{H}}{}_S^n$; if S is complete intersection, then it is Gorenstein and the exact sequence yields equalities. A closer look at the numbers delivers the last assertion in (3).

In the arguments above the exact sequence in Theorem C.5.10 was used only at the level of vector spaces. By contrast, the construction of that sequence, which we sketch below, is powered by graded bimodule structures.

The map ι appears in the exact sequence (C.5.6) of graded H_S-modules when one chooses $M = k = N$. The third space in that sequence is $\overline{\mathrm{Ext}}_S(k,k)$; by (C.5.7), it is isomorphic to $\mathrm{Tor}^S(\mathrm{RHom}_S(k,S),k)$, and hence to $\mathrm{Ext}_S(k,S)[1]\otimes_k \mathrm{H}^S$ as graded vector spaces. In order to match the actions of H_S, it is more expedient to take a step back and note that the isomorphism

$$\mathrm{Hom}_S(\mathbf{P}(k),S)\otimes_S \mathbf{P}(k) \xrightarrow{\cong} \overline{\mathrm{Hom}}_S(\mathbf{P}(k),\mathbf{P}(k))$$

is one of left dg modules for the actions of $\mathrm{Hom}_S(\mathbf{P}(k),\mathbf{P}(k))$ on the right-hand copies of $\mathbf{P}(k)$.

Martsinkovsky [**97**] proved that ι is injective by showing that in (C.5.6) the map $\overline{\mathrm{Ext}}_S(k,k) \to \mathrm{H}_S$ is zero. As noted in ((C.5.8)), its source is an $\mathrm{H}_S^{\geqslant 1}$-torsion module, so it exhibits a non-zero-divisor in $\mathrm{H}_S^{\geqslant 1}$. This can be done whenever S is not regular, as $\widehat{\mathrm{H}}_S$ is the universal enveloping algebra of a positively graded super-Lie-algebra π_S with $\pi_S^2 \neq 0$. It remains to remark that the Poincaré–Birkhoff–Witt Theorem implies that no non-zero element in π_S of even degree is a zero-divisor; see [**10**, Section 10].

Theorem C.5.10 suggests to study the algebraic structure of the k-algebra $\widehat{\mathrm{H}}_S$ through the ring extension $\iota\colon \mathrm{H}_S \to \widehat{\mathrm{H}}_S$. In the simplest case, that of a *hypersurface ring* S, the result is a special case of Buchweitz's Theorem 10.2.3; for any pair (M,N) of S-modules, he identifies the map $\mathrm{Ext}_S(M,N) \to \widehat{\mathrm{Ext}}_S(M,N)$, induced by the stabilization functor $\mathrm{D^b}(S) \to \underline{\mathrm{D}}^{\mathrm{b}}(S)$, as the inversion of "the" Gulliksen–Eisenbud operator.

The next simplest case, when S is complete intersection of codimension $c \geq 2$, exhibits a very different behavior. To describe it, we consider H_S as a subalgebra $\widehat{\mathrm{H}}_S$ via the homomorphism ι and set

$$T_S := \{t \in \widehat{\mathrm{H}}_S \mid \mathrm{H}^{\geq i}\, t = 0 \text{ for some } i\}\,.$$

This a two-sided ideal in the ring $\widehat{\mathrm{H}}_S$, there is an isomorphism $T_S \cong \mathrm{H}^S[1-d]$ as graded left H_S-modules, and equalities $\widehat{\mathrm{H}}_S = \mathrm{H}_S \oplus T_S$ such that $T_S\cdot T_S = 0$. We do not know if $\widehat{\mathrm{H}}_S$ is completely defined by T_S, because it is not clear to us if the isomorphism $T_S \cong \mathrm{H}^S[1-d]$ is also one of right H_S-modules.

These results extend the calculation of Tate cohomology of $\mathrm{Sym}(V)\otimes_K \bigwedge(W)$ given in Example 8.5.7 (iv). Buchweitz, Pham and Roberts clarified the situation in [**37**], where they explain how to compute the stable cohomology of a local complete intersection ring by reducing to the case of $\mathrm{Sym}(V)\otimes_K \bigwedge(W)$, using dg methods. They use this construction as well to produce universal complete resolutions over complete intersection rings. In [**18**] the multiplicative structure on $\widehat{\mathrm{Ext}}_R(k,k)$ is discussed also for the class of Gorenstein rings, and for Golod rings.

(C.5.11) *Artin algebras.* Outside the context of commutative rings, there are few results as clear-cut as those discussed in ((C.5.9)). That observation channels a feeling that cohomological structures over commutative rings are more "rigid". For instance, compare the structure theorem (C.5.10) to its prototype, a theorem of Benson and Carlson [**28**] about the cohomology of a finite group G with coefficients in a field k whose characteristic divides $|G|$: If the cohomology ring $\mathrm{H}^*(G,k)$ has depth at least 2, then in the Tate cohomology ring $\widehat{\mathrm{H}}^*(G,k)$ the product of any two elements of negative degree is zero.

There are a number of publications dealing with stable cohomology for Artin algebras, and also for not necessarily Iwanaga–Gorenstein algebras. One encounters new phenomena in this context, which provide a counterpoint to the behavior of stable cohomology over commutative rings. For instance, it follows from results in Chapter 4 and Happel [**65**] that if S,T are noetherian rings that are derived equivalent, then there is an equivalence $\underline{\mathrm{D}}^{\mathrm{b}}(S) \cong \underline{\mathrm{D}}^{\mathrm{b}}(T)$ of triangulated categories.

On the other hand, Chen [**44**] describes a procedure to construct an Artinian ring that is not Gorenstein but whose stable derived category is equivalent to the stable derived category of a Gorenstein Artinian ring. This shows in particular that finiteness of the module of morphisms in the stable derived category does not detect the Gorenstein property, in contrast with situation over commutative rings discussed in ((C.5.9)).

Chen [**45**] studies the stable cohomology of an Artin algebra of radical square zero, and gives criteria for the Hom-finiteness of the stable derived category in terms of an associated quiver. When the quiver is a point this yields a calculation of the stable cohomology of the trivial extension algebra $A := k \ltimes V$, where V is a finite dimensional vector space over a field k; see also [**118**]. This A is an example of a commutative Golod ring and this calculation can also be recovered from [**18**, Section 10]. We refer to [**1**] and the references therein for other results concerning stable module categories of Artin algebras.

C.6 Annihilator of stable derived categories. Next we discuss annihilators of the stable category, a topic that Buchweitz considers in 7.8. Let R be a commutative ring and S an R-algebra, not necessarily commutative. For any X,Y in $\underline{\mathrm{D}}^{\mathrm{b}}(S)$, the ring R acts on $\widehat{\mathrm{Ext}}_S(X,Y)$ via multiplication on either component. The *annihilator* of $\underline{\mathrm{D}}^{\mathrm{b}}(S)$ consists of the elements of R that annihilate $\widehat{\mathrm{Ext}}_S(X,Y)$ for all X,Y; we denote it $\mathrm{ann}_R\,\underline{\mathrm{D}}^{\mathrm{b}}(S)$. When S is strongly Gorenstein in the sense of 4.1, then by Theorem 4.4.1 and Corollary 6.4.1, the annihilator can be expressed in terms of maximal Cohen–Macaulay modules as follows:

$$\text{(C.6.1)}\qquad \mathrm{ann}_R\,\underline{\mathrm{D}}^{\mathrm{b}}(S) = \bigcap_{M,N\in \mathrm{MCM}(S)} \mathrm{ann}_R\,\mathrm{Ext}^1_S(M,N)\,.$$

Buchweitz views the annihilator of $\underline{\mathrm{D}}^{\mathrm{b}}(S)$ over its center $\mathrm{Z}(S)$, or rather, the closed subset of $\operatorname{Spec}\mathrm{Z}(S)$ it defines, as the "universal" support of Tate cohomology of S. It is one measure of the singularity of S; see further below. Computing the annihilator exactly is a difficult task in general, but there are certain canonical elements lying in it. One such is the (classical) Noether different of S over R, introduced in [**105**]; namely

$$\vartheta^0_N(S/R) := \mathrm{Im}(\mathrm{Hom}_{S^{\mathrm{e}}_R}(S,S^{\mathrm{e}}_R) \to \mathrm{Hom}_{S^{\mathrm{e}}_R}(S,S)) \subseteq \mathrm{Z}(S)$$

where $S^{\mathrm{e}}_R := S\otimes_R S^{\mathrm{op}}$ is the enveloping algebra of S over R. See also [**4**] and [**117**]. In Theorem 7.8.3 Buchweitz proves that when R is regular and S is strongly

Gorenstein and finitely generated projective as an R-module, one has an inclusion $\vartheta^0_N(S/R) \subseteq \operatorname{ann}_{Z(S)} \underline{\mathrm{D}}^{\mathrm{b}}(S)$. He also notes at the end of Section 7.8 that this inclusion is typically strict, even for commutative Gorenstein rings over fields.

Theorem 7.8.3 has been generalized to cover also the case where S is not Gorenstein, as follows: The derived enveloping algebra of S is

$$S^{(\mathrm{e})}_R := S \otimes^{\mathsf{L}}_R S^{\mathrm{op}},$$

and the *derived Noether different of S over R* is the graded R-module

$$\vartheta_{dN}(S/R) := \operatorname{Im}(\operatorname{Ext}_{S^{(\mathrm{e})}_R}(S, S^{(\mathrm{e})}_R) \to \operatorname{Ext}_{S^{(\mathrm{e})}_R}(S, S));$$

see [**77**, Section 2]. In degree 0 this is contained in the classical Noether different $\vartheta^0_N(S/R)$. Iyengar and Takahashi [**77**, Proposition 4.1] show that if S is any Noether algebra over a regular ring R, then the derived Noether different $\vartheta_{dN}(S/R)$ lies in the annihilator of $\underline{\mathrm{D}}^{\mathrm{b}}(S)$.

There are more refined results on the annihilator of the stable derived category when S is commutative, and we turn our attention to this context for the remainder of this discussion. When S is a one-dimensional reduced complete Gorenstein local ring, Esentepe proves that the annihilator is the conductor ideal of S, namely the ideal $(S :_S \overline{S}) = \operatorname{ann}_S(\overline{S}/S)$ where $\overline{S}$ is the integral closure of S inside its localization at the set of non-zero divisors [**58**, Theorem 4.3]. This is one of the few classes of rings for which a complete description of $\operatorname{ann}_S \underline{\mathrm{D}}^{\mathrm{b}}(S)$ is available. Esentepe's theorem extends an unpublished result of Buchweitz from the case of finite MCM representation type.

For general commutative rings, the classical Jacobian ideal is a source for elements in the annihilator, as we explain next. In [**77**, Section 3], also see [**126**], the Jacobian ideal of a general commutative noetherian ring S is defined to be

$$J(S) := \sum_{R \subseteq S} \vartheta_K(S/R)$$

where $\vartheta_K(S/R)$ the Kähler different, namely the 0th Fitting ideal of $\Omega_{S/R}$, the module of Kähler differentials of S over R; see [**91**, Section 10] and [**117**, §15]. The summation is over subrings R of S that are of finite global dimension, and such that S is a finitely generated R-module. For such an R, the Kähler different, the Noether different, and the derived Noether different of S over R are the same up to radical; see [**77**, Lemma 3.1], and also the discussion after Theorem 7.8.3. From this fact, and the discussion earlier, it follows that a power of $J(S)$ annihilates $\underline{\mathrm{D}}^{\mathrm{b}}(S)$; see [**77**, Theorem 4.2], where it is proved that under certain constraints on the depth of S, the entire Jacobian ideal annihilates $\underline{\mathrm{D}}^{\mathrm{b}}(S)$. This applies in particular when S is Cohen–Macaulay. If S is a finitely generated algebra over a field k, then $J(S)$ is the classical Jacobian ideal of S, namely, the dth Fitting ideal of $\Omega_{S/k}$, for $d = \dim S$; by [**126**, Theorem 2.3], see also [**117**].

It is also useful to consider uniform annihilators of Ext-modules of finitely generated modules, namely the ideal

$$\operatorname{ca}(S) := \bigcup_{n \geq 0} \operatorname{ann}_S \operatorname{Ext}^{\geqslant n}_S(-,-),$$

where $\operatorname{Ext}^i_S(-,-)$ is restricted to finitely generated S-modules. When S is a commutative Gorenstein ring this is the annihilator of $\underline{\mathrm{D}}^{\mathrm{b}}(S)$; see (C.6.1). The closed set defined by $\mathrm{ca}(S)$ is contained in $\mathrm{Sing}(S)$, the singular locus of S; this was observed already, though not cast in this language, in [**113, 125**]. It was proved in [**76**, Theorem 5.3 & 5.4] that $\mathrm{ca}(S)$ precisely defines the singular locus when S is an equicharacteristic excellent local ring of finite Krull dimension or when S is a (not necessarily local) ring essentially of finite type over a field.

When S is Cohen–Macaulay, the annihilator of $\mathrm{Ext}^1_S(-,-)$ has been studied among others by Dieterich [**50**], Popescu and Roczen [**113**], and Wang [**125, 126**]. Their work was inspired by the Brauer–Thrall conjectures over local Cohen–Macaulay rings. When S is a complete Cohen–Macaulay local ring containing a field and R is a normalization of S, then the Noether different $\vartheta^0_N(S/R)$ annihilates $\mathrm{Ext}^1_S(M,-)$ for any MCM module M; cf. [**125**, Corollary 5.13]. In particular, it lies in $\mathrm{ca}(S)$.

For any commutative noetherian ring S of Krull dimension d, a power of the Jacobian ideal $J(S)$ annihilates $\mathrm{Ext}^{d+1}_S(-,-)$; see [**77**, Theorem 3.4]. Annihilators of (stable) cohomology modules are connected to decompositions, and the existence of generators, for (stable) derived categories; see [**49, 75, 76**] for this connection, and also for further results on annihilator ideals.

C.7 Differential graded enhancements of singularity categories. So far we have not mentioned an important extra structure possessed by the stable derived category $\underline{\mathrm{D}}^{\mathrm{b}}(S)$: it is the homotopy category of a *pre-triangulated dg category*. In rough terms, this means that the class of morphisms in $\underline{\mathrm{D}}^{\mathrm{b}}(S)$ can be upgraded to form complexes, and that the composition operation can be lifted to this level as well; cf. [**81**].

A choice of such a pre-triangulated dg category is called an *enhancement* of $\underline{\mathrm{D}}^{\mathrm{b}}(S)$, and according to the work of Keller [**82**] there is a canonical enhancement compatible with the Verdier quotient $\mathrm{D}^{\mathrm{b}}(S) \to \underline{\mathrm{D}}^{\mathrm{b}}(S)$. This is made explicit in the work of Blanc, Robalo, Toen, and Vezzosi [**29**], where a connection is made with the theory of vanishing cycles.

In general, the presence of an enhancement allows one to compute important invariants of the algebra S which are not otherwise available. This is exploited in the work of Brown and Dyckerhoff [**34**] to compute topological K-theory. More generally Gratz and Stevenson explain in [**62**] how to compute (for certain rings) any $\mathbb{A}^1$-homotopy invariant of the enhanced stable derived category, by building on work of Tabuada [**121**]. Keller [**83**] proved that the Tate–Hochschild cohomology of an algebra can be computed from its enhanced stable derived category.

Bibliography

[AB69] Maurice Auslander and Mark Bridger, *Stable module theory*, Memoirs of the American Mathematical Society, No. 94, American Mathematical Society, Providence, R.I., 1969. MR0269685

[AGV72] P. Deligne, *Cohomologie étale* (French), Lecture Notes in Mathematics, vol. 569, Springer-Verlag, Berlin, 1977. Séminaire de géométrie algébrique du Bois-Marie SGA $4\frac{1}{2}$, DOI 10.1007/BFb0091526. MR463174

[67] *Anneaux de Gorenstein, et torsion en algèbre commutative* (French), École Normale Supérieure de Jeunes Filles. Secrétariat mathématique, Paris, 1967. Séminaire d'Algèbre Commutative dirigé par Pierre Samuel, 1966/67; Texte rédigé, d'après des exposés de Maurice Auslander, Marguerite Mangeney, Christian Peskine et Lucien Szpiro. MR0225844

[Aus86] Maurice Auslander, *Isolated singularities and existence of almost split sequences*, Representation theory, II (Ottawa, Ont., 1984), Lecture Notes in Math., vol. 1178, Springer, Berlin, 1986, pp. 194–242, DOI 10.1007/BFb0075297. MR842486

[Avr84] Luchezar L. Avramov, *Local algebra and rational homotopy*, Algebraic homotopy and local algebra (Luminy, 1982), Astérisque, vol. 113, Soc. Math. France, Paris, 1984, pp. 15–43. MR749041

[Bas68] Hyman Bass, *Algebraic K-theory*, W. A. Benjamin, Inc., New York-Amsterdam, 1968. MR0249491

[BBD82] A. A. Beĭlinson, J. Bernstein, and P. Deligne, *Faisceaux pervers* (French), Analysis and topology on singular spaces, I (Luminy, 1981), Astérisque, vol. 100, Soc. Math. France, Paris, 1982, pp. 5–171. MR751966

[BEH87] Ragnar-Olaf Buchweitz, David Eisenbud, and Jürgen Herzog, *Cohen-Macaulay modules on quadrics*, Singularities, representation of algebras, and vector bundles (Lambrecht, 1985), Lecture Notes in Math., vol. 1273, Springer, Berlin, 1987, pp. 58–116, DOI 10.1007/BFb0078838. MR915169

[Beĭ84] A. A. Beĭlinson, *The derived category of coherent sheaves on $\mathbf{P}^n$*, Selecta Math. Soviet. **3** (1983/84), no. 3, 233–237. Selected translations. MR863137

[Ber61] Robert Berger, *Über verschiedene Differentenbegriffe* (German), S.-B. Heidelberger Akad. Wiss. Math.-Nat. Kl. **1960/61** (1960/1961), 1–44. MR0131443

[BGG78] I. N. Bernšteĭn, I. M. Gel′fand, and S. I. Gel′fand, *Algebraic vector bundles on $\mathbf{P}^n$ and problems of linear algebra* (Russian), Funktsional. Anal. i Prilozhen. **12** (1978), no. 3, 66–67. MR509387

[BGI71] *Revêtements étales et groupe fondamental* (French), Lecture Notes in Mathematics, Vol. 224, Springer-Verlag, Berlin-New York, 1971. Séminaire de Géométrie Algébrique du Bois Marie 1960–1961 (SGA 1); Dirigé par Alexandre Grothendieck. Augmenté de deux exposés de M. Raynaud. MR0354651

[BGS87] R.-O. Buchweitz, G.-M. Greuel, and F.-O. Schreyer, *Cohen-Macaulay modules on hypersurface singularities. II*, Invent. Math. **88** (1987), no. 1, 165–182, DOI 10.1007/BF01405096. MR877011

[Bjö79] J.-E. Björk, *Rings of differential operators*, North-Holland Mathematical Library, vol. 21, North-Holland Publishing Co., Amsterdam-New York, 1979. MR549189

[Bou70] N. Bourbaki, *Éléments de mathématique. Algèbre. Chapitres 1 à 3* (French), Hermann, Paris, 1970. MR0274237

[Bou80] ———, *Éléments de mathématique*, Masson, Paris, 1980, Algèbre. Chapitre 10. Algèbre homologique. [Algebra. Chapter 10. Homological algebra]. MR610795

[BR86] Jörgen Backelin and Jan-Erik Roos, *When is the double Yoneda Ext-algebra of a local Noetherian ring again Noetherian?*, Algebra, algebraic topology and their interactions (Stockholm, 1983), Lecture Notes in Math., vol. 1183, Springer, Berlin, 1986, pp. 101–119, DOI 10.1007/BFb0075452. MR846441

[Bro82] Kenneth S. Brown, *Cohomology of groups*, Graduate Texts in Mathematics, vol. 87, Springer-Verlag, New York-Berlin, 1982. MR672956

[Car86] Gunnar Carlsson, *Free* $(\mathbf{Z}/2)^k$*-actions and a problem in commutative algebra*, Transformation groups, Poznań 1985, Lecture Notes in Math., vol. 1217, Springer, Berlin, 1986, pp. 79–83, DOI 10.1007/BFb0072815. MR874170

[CE99] Henri Cartan and Samuel Eilenberg, *Homological algebra*, Princeton Landmarks in Mathematics, Princeton University Press, Princeton, NJ, 1999. With an appendix by David A. Buchsbaum; Reprint of the 1956 original. MR1731415

[CR81] Charles W. Curtis and Irving Reiner, *Methods of representation theory. Vol. I*, Pure and Applied Mathematics, John Wiley & Sons, Inc., New York, 1981. With applications to finite groups and orders. MR632548

[Eis80] David Eisenbud, *Homological algebra on a complete intersection, with an application to group representations*, Trans. Amer. Math. Soc. **260** (1980), no. 1, 35–64, DOI 10.2307/1999875. MR570778

[FGR75] Robert M. Fossum, Phillip A. Griffith, and Idun Reiten, *Trivial extensions of abelian categories*, Lecture Notes in Mathematics, Vol. 456, Springer-Verlag, Berlin-New York, 1975. Homological algebra of trivial extensions of abelian categories with applications to ring theory. MR0389981

[FHJ+88] Yves Félix, Stephen Halperin, Carl Jacobsson, Clas Löfwall, and Jean-Claude Thomas, *The radical of the homotopy Lie algebra*, Amer. J. Math. **110** (1988), no. 2, 301–322, DOI 10.2307/2374504. MR935009

[Fos66] Robert Fossum, *The Noetherian different of projective orders*, J. Reine Angew. Math. **224** (1966), 207–218, DOI 10.1515/crll.1966.224.207. MR222067

[Gel84] S. I. Gelfand, *Appendix to the russian translation of* [**OSS11**], "Mir", Moscow, 1984.

[Gro66] A. Grothendieck, *Éléments de géométrie algébrique. IV. Étude locale des schémas et des morphismes de schémas. III*, Inst. Hautes Études Sci. Publ. Math. **28** (1966), 255. MR217086

[Gro67] ______, *Éléments de géométrie algébrique. IV. étude locale des schémas et des morphismes de schémas IV*, Inst. Hautes Études Sci. Publ. Math. (1967), no. 32, 361. MR0238860

[Hap85] Dieter Happel, *Triangulated categories in the representation theory of finite-dimensional algebras*, London Mathematical Society Lecture Note Series, vol. 119, Cambridge University Press, Cambridge, 1988, DOI 10.1017/CBO9780511629228. MR935124

[Har66] Robin Hartshorne, *Residues and duality*, Lecture Notes in Mathematics, No. 20, Springer-Verlag, Berlin-New York, 1966. Lecture notes of a seminar on the work of A. Grothendieck, given at Harvard 1963/64; With an appendix by P. Deligne. MR0222093

[Hel60] Alex Heller, *The loop-space functor in homological algebra*, Trans. Amer. Math. Soc. **96** (1960), 382–394, DOI 10.2307/1993530. MR116045

[Her78] Jürgen Herzog, *Ringe mit nur endlich vielen Isomorphieklassen von maximalen, unzerlegbaren Cohen-Macaulay-Moduln* (German), Math. Ann. **233** (1978), no. 1, 21–34, DOI 10.1007/BF01351494. MR463155

[HK71] Ernst Kunz, *Komplementärmodul und kanonischer Modul* (German), Der kanonische Modul eines Cohen-Macaulay-Rings (Sem. Lokale Kohomologietheorie von Grothendieck, Univ. Regensburg, Regensburg, 1970/1971), Springer, Berlin, 1971, pp. 85–102, 103. Lecture Notes in Math., Vol. 238. MR0476717

[Ill71] Luc Illusie, *Complexe cotangent et déformations. I* (French), Lecture Notes in Mathematics, Vol. 239, Springer-Verlag, Berlin-New York, 1971. MR0491680

[Ill77] *Cohomologie l-adique et fonctions L* (French), Lecture Notes in Mathematics, Vol. 589, Springer-Verlag, Berlin-New York, 1977. Séminaire de Géometrie Algébrique du Bois-Marie 1965–1966 (SGA 5); Edité par Luc Illusie. MR0491704

[Kap74] Irving Kaplansky, *Commutative rings*, Revised edition, The University of Chicago Press, Chicago, Ill.-London, 1974. MR0345945

[Kle80] Steven L. Kleiman, *Relative duality for quasicoherent sheaves*, Compositio Math. **41** (1980), no. 1, 39–60. MR578050

[Knö87] Horst Knörrer, *Cohen-Macaulay modules on hypersurface singularities. I*, Invent. Math. **88** (1987), no. 1, 153–164, DOI 10.1007/BF01405095. MR877010

[Kun68] Ernst Kunz, *Vollständige Durchschnitte und Differenten* (German), Arch. Math. (Basel) **19** (1968), 47–58, DOI 10.1007/BF01898801. MR224603

[Lei83] Jürgen Leiterer, *Subsheaves in bundles on* $\mathbf{P}_n$ *and the Penrose transform*, Analytic functions, Błażejewko 1982 (Błażejewko, 1982), Lecture Notes in Math., vol. 1039, Springer, Berlin, 1983, pp. 332–345, DOI 10.1007/BFb0073375. MR739354

[Löf86] Clas Löfwall, *On the subalgebra generated by the one-dimensional elements in the Yoneda Ext-algebra*, Algebra, algebraic topology and their interactions (Stockholm, 1983), Lecture Notes in Math., vol. 1183, Springer, Berlin, 1986, pp. 291–338, DOI 10.1007/BFb0075468. MR846457

[OP84] O. V. Ogievetskiĭ and I. B. Penkov, *Serre duality for projective supermanifolds* (Russian), Funktsional. Anal. i Prilozhen. **18** (1984), no. 1, 78–79. MR739100

[OSS11] Christian Okonek, Michael Schneider, and Heinz Spindler, *Vector bundles on complex projective spaces*, Modern Birkhäuser Classics, Birkhäuser/Springer Basel AG, Basel, 2011. Corrected reprint of the 1988 edition; With an appendix by S. I. Gelfand. MR2815674

[Qui69] Daniel Quillen, *Rational homotopy theory*, Ann. of Math. (2) **90** (1969), 205–295, DOI 10.2307/1970725. MR258031

[Qui73] Daniel Quillen, *Higher algebraic K-theory. I*, Algebraic K-theory, I: Higher K-theories (Proc. Conf., Battelle Memorial Inst., Seattle, Wash., 1972), Springer, Berlin, 1973, pp. 85–147. Lecture Notes in Math., Vol. 341. MR0338129

[Roo73] Jan-Erik Roos, *Compléments à l'étude des quotients primitifs des algèbres enveloppantes des algèbres de Lie semi-simples* (French), C. R. Acad. Sci. Paris Sér. A-B **276** (1973), A447–A450. MR318250

[Ser79] Jean-Pierre Serre, *Local fields*, Graduate Texts in Mathematics, vol. 67, Springer-Verlag, New York-Berlin, 1979. Translated from the French by Marvin Jay Greenberg. MR554237

[Ser02] Jean-Pierre Serre, *Galois cohomology*, Corrected reprint of the 1997 English edition, Springer Monographs in Mathematics, Springer-Verlag, Berlin, 2002. Translated from the French by Patrick Ion and revised by the author. MR1867431

[Slo80] Peter Slodowy, *Simple singularities and simple algebraic groups*, Lecture Notes in Mathematics, vol. 815, Springer, Berlin, 1980. MR584445

[Sta82] J. T. Stafford, *Homological properties of the enveloping algebra* $U(\mathrm{Sl}_2)$, Math. Proc. Cambridge Philos. Soc. **91** (1982), no. 1, 29–37, DOI 10.1017/S0305004100059089. MR633253

[Swa68] R. G. Swan, *Algebraic K-theory*, Lecture Notes in Mathematics, No. 76, Springer-Verlag, Berlin-New York, 1968. MR0245634

[Swa85] Richard G. Swan, *K-theory of quadric hypersurfaces*, Ann. of Math. (2) **122** (1985), no. 1, 113–153, DOI 10.2307/1971371. MR799254

[Ver77] Jean-Louis Verdier, *Catégories dérivées: quelques résultats (état 0)* (French), Cohomologie étale, Lecture Notes in Math., vol. 569, Springer, Berlin, 1977, pp. 262–311. MR3727440

[Zak69] Abraham Zaks, *Injective dimension of semi-primary rings*, J. Algebra **13** (1969), 73–86, DOI 10.1016/0021-8693(69)90007-6. MR244325

Additional bibliography

[1] Claire Amiot, Osamu Iyama, and Idun Reiten, *Stable categories of Cohen-Macaulay modules and cluster categories*, Amer. J. Math. **137** (2015), no. 3, 813–857, DOI 10.1353/ajm.2015.0019. MR3357123

[2] M. Artin and J. J. Zhang, *Noncommutative projective schemes*, Adv. Math. **109** (1994), no. 2, 228–287, DOI 10.1006/aima.1994.1087. MR1304753

[3] Maurice Auslander, *Functors and morphisms determined by objects*, Representation theory of algebras (Proc. Conf., Temple Univ., Philadelphia, Pa., 1976), Dekker, New York, 1978, pp. 1–244. Lecture Notes in Pure Appl. Math., Vol. 37. MR0480688

[4] M. Auslander and D. A. Buchsbaum, *On ramification theory in noetherian rings*, Amer. J. Math. **81** (1959), 749–765, DOI 10.2307/2372926. MR106929

[5] Maurice Auslander and Ragnar-Olaf Buchweitz, *The homological theory of maximal Cohen-Macaulay approximations* (English, with French summary), Mém. Soc. Math. France (N.S.) **38** (1989), 5–37. Colloque en l'honneur de Pierre Samuel (Orsay, 1987). MR1044344

[6] Maurice Auslander and Idun Reiten, *Almost split sequences for* **Z***-graded rings*, Singularities, representation of algebras, and vector bundles (Lambrecht, 1985), Lecture Notes in Math., vol. 1273, Springer, Berlin, 1987, pp. 232–243, DOI 10.1007/BFb0078847. MR915178

[7] Maurice Auslander and Idun Reiten, *The Cohen-Macaulay type of Cohen-Macaulay rings*, Adv. in Math. **73** (1989), no. 1, 1–23, DOI 10.1016/0001-8708(89)90057-1. MR979585

[8] Maurice Auslander and Idun Reiten, *Cohen-Macaulay and Gorenstein Artin algebras*, Representation theory of finite groups and finite-dimensional algebras (Bielefeld, 1991), Progr. Math., vol. 95, Birkhäuser, Basel, 1991, pp. 221–245. MR1112162

[9] Maurice Auslander, Idun Reiten, and Sverre O. Smalø, *Representation theory of Artin algebras*, Cambridge Studies in Advanced Mathematics, vol. 36, Cambridge University Press, Cambridge, 1997. Corrected reprint of the 1995 original. MR1476671

[10] Luchezar L. Avramov, *Infinite free resolutions*, Six lectures on commutative algebra (Bellaterra, 1996), Progr. Math., vol. 166, Birkhäuser, Basel, 1998, pp. 1–118. MR1648664

[11] Luchezar L. Avramov and Hans-Bjørn Foxby, *Homological dimensions of unbounded complexes*, J. Pure Appl. Algebra **71** (1991), no. 2-3, 129–155, DOI 10.1016/0022-4049(91)90144-Q. MR1117631

[12] Luchezar L. Avramov, Vesselin N. Gasharov, and Irena V. Peeva, *Complete intersection dimension*, Inst. Hautes Études Sci. Publ. Math. **86** (1997), 67–114 (1998). MR1608565

[13] Luchezar L. Avramov and Srikanth B. Iyengar, *Bass numbers over local rings via stable cohomology*, J. Commut. Algebra **5** (2013), no. 1, 5–15, DOI 10.1216/jca-2013-5-1-5. MR3084119

[14] Luchezar L. Avramov, Srikanth B. Iyengar, and Joseph Lipman, *Reflexivity and rigidity for complexes. I. Commutative rings*, Algebra Number Theory **4** (2010), no. 1, 47–86, DOI 10.2140/ant.2010.4.47. MR2592013

[15] Luchezar L. Avramov, Srikanth B. Iyengar, and Liana M. Şega, *Free resolutions over short local rings*, J. Lond. Math. Soc. (2) **78** (2008), no. 2, 459–476, DOI 10.1112/jlms/jdn027. MR2439635

[16] Luchezar L. Avramov, Andrew R. Kustin, and Matthew Miller, *Poincaré series of modules over local rings of small embedding codepth or small linking number*, J. Algebra **118** (1988), no. 1, 162–204, DOI 10.1016/0021-8693(88)90056-7. MR961334

[17] Luchezar L. Avramov and Alex Martsinkovsky, *Absolute, relative, and Tate cohomology of modules of finite Gorenstein dimension*, Proc. London Math. Soc. (3) **85** (2002), no. 2, 393–440, DOI 10.1112/S0024611502013527. MR1912056

[18] Luchezar L. Avramov and Oana Veliche, *Stable cohomology over local rings*, Adv. Math. **213** (2007), no. 1, 93–139, DOI 10.1016/j.aim.2006.11.012. MR2331239

[19] Matthew Ballard, David Favero, and Ludmil Katzarkov, *Orlov spectra: bounds and gaps*, Invent. Math. **189** (2012), no. 2, 359–430, DOI 10.1007/s00222-011-0367-y. MR2947547

[20] Matthew Ballard, David Favero, and Ludmil Katzarkov, *Variation of geometric invariant theory quotients and derived categories*, J. Reine Angew. Math. **746** (2019), 235–303, DOI 10.1515/crelle-2015-0096. MR3895631

[21] Paul Balmer, Ivo Dell'Ambrogio, and Beren Sanders, *Grothendieck-Neeman duality and the Wirthmüller isomorphism*, Compos. Math. **152** (2016), no. 8, 1740–1776, DOI 10.1112/S0010437X16007375. MR3542492

[22] Vladimir Baranovsky and Jeremy Pecharich, *On equivalences of derived and singular categories*, Cent. Eur. J. Math. **8** (2010), no. 1, 1–14, DOI 10.2478/s11533-009-0063-y. MR2593258

[23] Hyman Bass, *On the ubiquity of Gorenstein rings*, Math. Z. **82** (1963), 8–28, DOI 10.1007/BF01112819. MR153708

[24] Alexander Beilinson, Victor Ginzburg, and Wolfgang Soergel, *Koszul duality patterns in representation theory*, J. Amer. Math. Soc. **9** (1996), no. 2, 473–527, DOI 10.1090/S0894-0347-96-00192-0. MR1322847

[25] A. A. Beĭlinson, *Coherent sheaves on* $\mathbf{P}^n$ *and problems in linear algebra* (Russian), Funktsional. Anal. i Prilozhen. **12** (1978), no. 3, 68–69. MR509388

[26] Apostolos Beligiannis, *The homological theory of contravariantly finite subcategories: Auslander-Buchweitz contexts, Gorenstein categories and (co-)stabilization*, Comm. Algebra **28** (2000), no. 10, 4547–4596, DOI 10.1080/00927870008827105. MR1780017

[27] Dave Benson, Srikanth B. Iyengar, Henning Krause, and Julia Pevtsova, *Local duality for representations of finite group schemes*, Compos. Math. **155** (2019), no. 2, 424–453, DOI 10.1112/s0010437x19007061. MR3914144

[28] D. J. Benson and Jon F. Carlson, *Products in negative cohomology*, J. Pure Appl. Algebra **82** (1992), no. 2, 107–129, DOI 10.1016/0022-4049(92)90116-W. MR1182934

[29] Anthony Blanc, Marco Robalo, Bertrand Toën, and Gabriele Vezzosi, *Motivic realizations of singularity categories and vanishing cycles* (English, with English and French summaries), J. Éc. polytech. Math. **5** (2018), 651–747, DOI 10.5802/jep.81. MR3877165

[30] Rikard Bøgvad and Stephen Halperin, *On a conjecture of Roos*, Algebra, algebraic topology and their interactions (Stockholm, 1983), Lecture Notes in Math., vol. 1183, Springer, Berlin, 1986, pp. 120–127, DOI 10.1007/BFb0075453. MR846442

[31] A. I. Bondal and M. M. Kapranov, *Representable functors, Serre functors, and reconstructions* (Russian), Izv. Akad. Nauk SSSR Ser. Mat. **53** (1989), no. 6, 1183–1205, 1337, DOI 10.1070/IM1990v035n03ABEH000716; English transl., Math. USSR-Izv. **35** (1990), no. 3, 519–541. MR1039961

[32] Michel Broué, *Equivalences of blocks of group algebras*, Finite-dimensional algebras and related topics (Ottawa, ON, 1992), NATO Adv. Sci. Inst. Ser. C Math. Phys. Sci., vol. 424, Kluwer Acad. Publ., Dordrecht, 1994, pp. 1–26, DOI 10.1007/978-94-017-1556-0_1. MR1308978

[33] Michael K. Brown, *Knörrer periodicity and Bott periodicity*, Doc. Math. **21** (2016), 1459–1501, DOI 10.2118/180920-pa. MR3603927

[34] Michael K. Brown and Tobias Dyckerhoff, *Topological K-theory of equivariant singularity categories*, Homology Homotopy Appl. **22** (2020), no. 2, 1–29, DOI 10.4310/hha.2020.v22.n2.a1. MR4073075

[35] R.-O. Buchweitz, *Maximal Cohen-Macaulay modules and Orlov's theorem, I*, Recorded Talk, 2012, Workshop 'Linking representation theory, singularity theory and non-commutative algebraic geometry' at Banff International Research Station.

[36] R.-O. Buchweitz, *Maximal Cohen-Macaulay modules and Orlov's theorem, II*, Recorded Talk, 2012, Workshop 'Linking representation theory, singularity theory and non-commutative algebraic geometry' at Banff International Research Station.

[37] *Representation theory of quivers and finite dimensional algebras*, Oberwolfach Rep. **8** (2011), no. 1, 523–608, DOI 10.4171/OWR/2011/10. Abstracts from the workshop held February 20–26, 2011; Organized by Bill Crawley-Boevey, Bernhard Keller, Henning Krause and Oyvind Solberg; Oberwolfach Reports. Vol. 8, no. 1. MR2849508

[38] Ragnar-Olaf Buchweitz and Duco Van Straten, *An index theorem for modules on a hypersurface singularity* (English, with English and Russian summaries), Mosc. Math. J. **12** (2012), no. 2, 237–259, 459, DOI 10.17323/1609-4514-2012-12-2-237-259. MR2978754

[39] Theo Bühler, *Exact categories*, Expo. Math. **28** (2010), no. 1, 1–69, DOI 10.1016/j.exmath.2009.04.004. MR2606234

[40] Igor Burban and Martin Kalck, *The relative singularity category of a non-commutative resolution of singularities*, Adv. Math. **231** (2012), no. 1, 414–435, DOI 10.1016/j.aim.2012.05.012. MR2935394

[41] Jesse Burke and Greg Stevenson, *The derived category of a graded Gorenstein ring*, Commutative algebra and noncommutative algebraic geometry. Vol. II, Math. Sci. Res. Inst. Publ., vol. 68, Cambridge Univ. Press, New York, 2015, pp. 93–123. MR3496862

[42] Jesse Burke and Mark E. Walker, *Matrix factorizations in higher codimension*, Trans. Amer. Math. Soc. **367** (2015), no. 5, 3323–3370, DOI 10.1090/S0002-9947-2015-06323-5. MR3314810

[43] Henri Cartan, *Œuvres. Vol. I, II, III* (French), Springer-Verlag, Berlin-New York, 1979. Edited by Reinhold Remmert and Jean-Pierre Serre. MR540747

[44] Xiao-Wu Chen, *Singularity categories, Schur functors and triangular matrix rings*, Algebr. Represent. Theory **12** (2009), no. 2-5, 181–191, DOI 10.1007/s10468-009-9149-2. MR2501179

[45] Xiao-Wu Chen, *The singularity category of an algebra with radical square zero*, Doc. Math. **16** (2011), 921–936. MR2880676

[46] Lars Winther Christensen, Greg Piepmeyer, Janet Striuli, and Ryo Takahashi, *Finite Gorenstein representation type implies simple singularity*, Adv. Math. **218** (2008), no. 4, 1012–1026, DOI 10.1016/j.aim.2008.03.005. MR2419377

[47] Joel M. Cohen, *Coherent graded rings and the non-existence of spaces of finite stable homotopy type*, Comment. Math. Helv. **44** (1969), 217–228, DOI 10.1007/BF02564524. MR247628

[48] Aldo Conca, Srikanth B. Iyengar, Hop D. Nguyen, and Tim Römer, *Absolutely Koszul algebras and the Backelin-Roos property*, Acta Math. Vietnam. **40** (2015), no. 3, 353–374, DOI 10.1007/s40306-015-0125-0. MR3395644

[49] Hailong Dao and Ryo Takahashi, *The radius of a subcategory of modules*, Algebra Number Theory **8** (2014), no. 1, 141–172, DOI 10.2140/ant.2014.8.141. MR3207581

[50] Ernst Dieterich, *Reduction of isolated singularities*, Comment. Math. Helv. **62** (1987), no. 4, 654–676, DOI 10.1007/BF02564469. MR920064

[51] Christopher M. Drupieski and Jonathan R. Kujawa, *Support varieties and modules of finite projective dimension for modular Lie superalgebras*, Algebra Number Theory **15** (2021), no. 5, 1157–1180, DOI 10.2140/ant.2021.15.1157. With an appendix by Luchezar L. Avramov and Srikanth B. Iyengar. MR4283100

[52] W. G. Dwyer, J. P. C. Greenlees, and S. Iyengar, *Duality in algebra and topology*, Adv. Math. **200** (2006), no. 2, 357–402, DOI 10.1016/j.aim.2005.11.004. MR2200850

[53] Tobias Dyckerhoff and Daniel Murfet, *Pushing forward matrix factorizations*, Duke Math. J. **162** (2013), no. 7, 1249–1311, DOI 10.1215/00127094-2142641. MR3079249

[54] Alexander I. Efimov and Leonid Positselski, *Coherent analogues of matrix factorizations and relative singularity categories*, Algebra Number Theory **9** (2015), no. 5, 1159–1292, DOI 10.2140/ant.2015.9.1159. MR3366002

[55] David Eisenbud and Jürgen Herzog, *The classification of homogeneous Cohen-Macaulay rings of finite representation type*, Math. Ann. **280** (1988), no. 2, 347–352, DOI 10.1007/BF01456058. MR929541

[56] Ioannis Emmanouil and Olympia Talelli, *Finiteness criteria in Gorenstein homological algebra*, Trans. Amer. Math. Soc. **366** (2014), no. 12, 6329–6351, DOI 10.1090/S0002-9947-2014-06007-8. MR3267012

[57] Ioannis Emmanouil and Olympia Talelli, *Gorenstein dimension and group cohomology with group ring coefficients*, J. Lond. Math. Soc. (2) **97** (2018), no. 2, 306–324, DOI 10.1112/jlms.12103. MR3789849

[58] Özgür Esentepe, *The cohomology annihilator of a curve singularity*, J. Algebra **541** (2020), 359–379, DOI 10.1016/j.jalgebra.2019.08.037. MR4017410

[59] Vincent Gelinas, *Contributions to the Stable Derived Categories of Gorenstein Rings*, ProQuest LLC, Ann Arbor, MI, 2018. Thesis (Ph.D.)–University of Toronto (Canada). MR3885608

[60] François Goichot, *Homologie de Tate-Vogel équivariante* (French, with English summary), J. Pure Appl. Algebra **82** (1992), no. 1, 39–64, DOI 10.1016/0022-4049(92)90009-5. MR1181092

[61] Shiro Goto, *Vanishing of* $\operatorname{Ext}^i_A(M, A)$, J. Math. Kyoto Univ. **22** (1982/83), no. 3, 481–484, DOI 10.1215/kjm/1250521731. MR674605

[62] Sira Gratz and Greg Stevenson, *Homotopy invariants of singularity categories*, J. Pure Appl. Algebra **224** (2020), no. 12, 106433, 18, DOI 10.1016/j.jpaa.2020.106433. MR4103346

[63] J. P. C. Greenlees and Greg Stevenson, *Morita theory and singularity categories*, Adv. Math. **365** (2020), 107055, 51, DOI 10.1016/j.aim.2020.107055. MR4065715

[64] Daniel Halpern-Leistner, *The derived category of a GIT quotient*, J. Amer. Math. Soc. **28** (2015), no. 3, 871–912, DOI 10.1090/S0894-0347-2014-00815-8. MR3327537

[65] Dieter Happel, *Triangulated categories in the representation theory of finite-dimensional algebras*, London Mathematical Society Lecture Note Series, vol. 119, Cambridge University Press, Cambridge, 1988, DOI 10.1017/CBO9780511629228. MR935124

[66] M. Herbst, K. Hori, and D. Page, *Phases of* $N = 2$ *theories in* $1 + 1$ *dimensions with boundary*, arXiv e-prints, 2008, `arXiv:0803.2045v1`, pp. 1–265.

[67] Jürgen Herzog and Srikanth Iyengar, *Koszul modules*, J. Pure Appl. Algebra **201** (2005), no. 1-3, 154–188, DOI 10.1016/j.jpaa.2004.12.037. MR2158753

[68] Melvin Hochster, *The dimension of an intersection in an ambient hypersurface*, Algebraic geometry (Chicago, Ill., 1980), Lecture Notes in Math., vol. 862, Springer, Berlin-New York, 1981, pp. 93–106. MR644818

[69] D. Huybrechts, *Fourier-Mukai transforms in algebraic geometry*, Oxford Mathematical Monographs, The Clarendon Press, Oxford University Press, Oxford, 2006, DOI 10.1093/acprof:oso/9780199296866.001.0001. MR2244106

[70] Yasuo Iwanaga, *On rings with finite self-injective dimension*, Comm. Algebra **7** (1979), no. 4, 393–414, DOI 10.1080/00927877908822356. MR522552

[71] Yasuo Iwanaga, *On rings with finite self-injective dimension. II*, Tsukuba J. Math. **4** (1980), no. 1, 107–113, DOI 10.21099/tkbjm/1496158797. MR597688

[72] Osamu Iyama and Dong Yang, *Quotients of triangulated categories and equivalences of Buchweitz, Orlov, and Amiot-Guo-Keller*, Amer. J. Math. **142** (2020), no. 5, 1641–1659, DOI 10.1353/ajm.2020.0041. MR4150654

[73] S. B. Iyengar and H. Krause, *The Nakayama functor and its completion for Gorenstein algebras*, arXiv e-prints, 2020, `arXiv:2010.05676`, pp. 1–35.

[74] Srikanth B. Iyengar and Tim Römer, *Linearity defects of modules over commutative rings*, J. Algebra **322** (2009), no. 9, 3212–3237, DOI 10.1016/j.jalgebra.2009.04.023. MR2567417

[75] Srikanth B. Iyengar and Ryo Takahashi, *Annihilation of cohomology and decompositions of derived categories*, Homology Homotopy Appl. **16** (2014), no. 2, 231–237, DOI 10.4310/HHA.2014.v16.n2.a12. MR3263893

[76] Srikanth B. Iyengar and Ryo Takahashi, *Annihilation of cohomology and strong generation of module categories*, Int. Math. Res. Not. IMRN **2** (2016), 499–535, DOI 10.1093/imrn/rnv136. MR3493424

[77] Srikanth B. Iyengar and Ryo Takahashi, *The Jacobian ideal of a commutative ring and annihilators of cohomology*, J. Algebra **571** (2021), 280–296, DOI 10.1016/j.jalgebra.2018.07.034. MR4200721

[78] Carl Jacobsson, *Finitely presented graded Lie algebras and homomorphisms of local rings*, J. Pure Appl. Algebra **38** (1985), no. 2-3, 243–253, DOI 10.1016/0022-4049(85)90012-X. MR814180

[79] David A. Jorgensen and Liana M. Şega, *Independence of the total reflexivity conditions for modules*, Algebr. Represent. Theory **9** (2006), no. 2, 217–226, DOI 10.1007/s10468-005-0559-5. MR2238367

[80] Martin Kalck and Dong Yang, *Relative singularity categories I: Auslander resolutions*, Adv. Math. **301** (2016), 973–1021, DOI 10.1016/j.aim.2016.06.011. MR3539395

[81] Bernhard Keller, *Deriving DG categories*, Ann. Sci. École Norm. Sup. (4) **27** (1994), no. 1, 63–102. MR1258406

[82] Bernhard Keller, *On the cyclic homology of exact categories*, J. Pure Appl. Algebra **136** (1999), no. 1, 1–56, DOI 10.1016/S0022-4049(97)00152-7. MR1667558

[83] Bernhard Keller, *Singular Hochschild cohomology via the singularity category* (English, with English and French summaries), C. R. Math. Acad. Sci. Paris **356** (2018), no. 11-12, 1106–1111, DOI 10.1016/j.crma.2018.10.003. MR3907577

[84] Bernhard Keller, *Corrections to "Singular Hochschild cohomology via the singularity category" [C. R. Acad. Sci. Paris, Ser. I 356 (2018) 1106–1111]* (English, with English and French summaries), C. R. Math. Acad. Sci. Paris **357** (2019), no. 6, 533–536, DOI 10.1016/j.crma.2019.06.004. MR3983279

[85] Bernhard Keller, Daniel Murfet, and Michel Van den Bergh, *On two examples by Iyama and Yoshino*, Compos. Math. **147** (2011), no. 2, 591–612, DOI 10.1112/S0010437X10004902. MR2776613

[86] Bernhard Keller and Dieter Vossieck, *Sous les catégories dérivées* (French, with English summary), C. R. Acad. Sci. Paris Sér. I Math. **305** (1987), no. 6, 225–228. MR907948

[87] Henning Krause, *A short proof for Auslander's defect formula*, Linear Algebra Appl. **365** (2003), 267–270, DOI 10.1016/S0024-3795(02)00481-0. Special issue on linear algebra methods in representation theory. MR1987342

[88] Henning Krause, *The stable derived category of a Noetherian scheme*, Compos. Math. **141** (2005), no. 5, 1128–1162, DOI 10.1112/S0010437X05001375. MR2157133

[89] Henning Krause, *Auslander-Reiten duality for Grothendieck abelian categories*, Trans. Amer. Math. Soc. **371** (2019), no. 4, 2455–2472, DOI 10.1090/tran/7379. MR3896086

[90] Peter H. Kropholler, *Hierarchical decompositions, generalized Tate cohomology, and groups of type* $(\mathrm{FP})_\infty$, Combinatorial and geometric group theory (Edinburgh, 1993), London Math. Soc. Lecture Note Ser., vol. 204, Cambridge Univ. Press, Cambridge, 1995, pp. 190–216. MR1320283

[91] Ernst Kunz, *Kähler differentials*, Advanced Lectures in Mathematics, Friedr. Vieweg & Sohn, Braunschweig, 1986, DOI 10.1007/978-3-663-14074-0. MR864975

[92] Ernst Kunz, *Residues and duality for projective algebraic varieties*, University Lecture Series, vol. 47, American Mathematical Society, Providence, RI, 2008. With the assistance of and contributions by David A. Cox and Alicia Dickenstein, DOI 10.1090/ulect/047. MR2464546

[93] T. Y. Lam, *A first course in noncommutative rings*, 2nd ed., Graduate Texts in Mathematics, vol. 131, Springer-Verlag, New York, 2001, DOI 10.1007/978-1-4419-8616-0. MR1838439

[94] Helmut Lenzing, *Weighted projective lines and applications*, Representations of algebras and related topics, EMS Ser. Congr. Rep., Eur. Math. Soc., Zürich, 2011, pp. 153–187, DOI 10.4171/101-1/5. MR2931898

[95] Graham J. Leuschke and Roger Wiegand, *Cohen-Macaulay representations*, Mathematical Surveys and Monographs, vol. 181, American Mathematical Society, Providence, RI, 2012, DOI 10.1090/surv/181. MR2919145

[96] Alex Martsinkovsky, *New homological invariants for modules over local rings. I*, J. Pure Appl. Algebra **110** (1996), no. 1, 1–8, DOI 10.1016/0022-4049(95)00092-5. MR1390668

[97] Alex Martsinkovsky, *A remarkable property of the (co) syzygy modules of the residue field of a nonregular local ring*, J. Pure Appl. Algebra **110** (1996), no. 1, 9–13, DOI 10.1016/0022-4049(95)00122-0. MR1390669

[98] Hideyuki Matsumura, *Commutative ring theory*, 2nd ed., Cambridge Studies in Advanced Mathematics, vol. 8, Cambridge University Press, Cambridge, 1989. Translated from the Japanese by M. Reid. MR1011461

[99] Nadia Mazza and Peter Symonds, *The stable category and invertible modules for infinite groups*, Adv. Math. **358** (2019), 106853, 26, DOI 10.1016/j.aim.2019.106853. MR4021877

[100] Hiroyuki Minamoto, *A criterion for graded coherence of tensor algebras and applications to higher dimensional AR-theory*, Eur. J. Math. **4** (2018), no. 2, 612–621, DOI 10.1007/s40879-017-0202-0. MR3799159

[101] Guido Mislin, *Tate cohomology for arbitrary groups via satellites*, Topology Appl. **56** (1994), no. 3, 293–300, DOI 10.1016/0166-8641(94)90081-7. MR1269317

[102] Daniel Murfet, *Residues and duality for singularity categories of isolated Gorenstein singularities*, Compos. Math. **149** (2013), no. 12, 2071–2100, DOI 10.1112/S0010437X13007082. MR3143706

[103] Masayoshi Nagata, *Local rings*, Interscience Tracts in Pure and Applied Mathematics, No. 13, Interscience Publishers a division of John Wiley & Sons New York-London, 1962. MR0155856

[104] A. Neeman, *The categories $\mathcal{T}^c$ and $\mathcal{T}_c^b$ determine each other*, arXiv e-prints, 2018, arXiv:1806.06471, pp. 1–30.

[105] Emmy Noether, *Idealdifferentiation und Differente* (German), J. Reine Angew. Math. **188** (1950), 1–21, DOI 10.1515/crll.1950.188.1. MR38337

[106] Brita E. A. Nucinkis, *Complete cohomology for arbitrary rings using injectives*, J. Pure Appl. Algebra **131** (1998), no. 3, 297–318, DOI 10.1016/S0022-4049(97)00082-0. MR1637023

[107] D. O. Orlov, *Triangulated categories of singularities and D-branes in Landau-Ginzburg models* (Russian, with Russian summary), Tr. Mat. Inst. Steklova **246** (2004), no. Algebr. Geom. Metody, Svyazi i Prilozh., 240–262; English transl., Proc. Steklov Inst. Math. **3(246)** (2004), 227–248. MR2101296

[108] D. O. Orlov, *Triangulated categories of singularities, and equivalences between Landau-Ginzburg models* (Russian, with Russian summary), Mat. Sb. **197** (2006), no. 12, 117–132, DOI 10.1070/SM2006v197n12ABEH003824; English transl., Sb. Math. **197** (2006), no. 11-12, 1827–1840. MR2437083

[109] Dmitri Orlov, *Derived categories of coherent sheaves and triangulated categories of singularities*, Algebra, arithmetic, and geometry: in honor of Yu. I. Manin. Vol. II, Progr. Math., vol. 270, Birkhäuser Boston, Boston, MA, 2009, pp. 503–531, DOI 10.1007/978-0-8176-4747-6_16. MR2641200

[110] Takuo Fukuda, *René Thom, la théorie des singularités et moi* (French), René Thom: portrait mathématique et philosophique, CNRS Éd., Paris, 2018, pp. 185–199. MR3888271

[111] Alexander Pavlov, *Betti tables of MCM modules over the cone of a plane cubic*, Math. Z. **297** (2021), no. 1-2, 223–254, DOI 10.1007/s00209-020-02509-5. MR4204691

[112] Dmitri Piontkovski, *Coherent algebras and noncommutative projective lines*, J. Algebra **319** (2008), no. 8, 3280–3290, DOI 10.1016/j.jalgebra.2007.07.010. MR2408318

[113] Dorin Popescu and Marko Roczen, *Indecomposable Cohen-Macaulay modules and irreducible maps*, Compositio Math. **76** (1990), no. 1-2, 277–294. Algebraic geometry (Berlin, 1988). MR1078867

[114] I. Reiten and M. Van den Bergh, *Noetherian hereditary abelian categories satisfying Serre duality*, J. Amer. Math. Soc. **15** (2002), no. 2, 295–366, DOI 10.1090/S0894-0347-02-00387-9. MR1887637

[115] Jeremy Rickard, *Morita theory for derived categories*, J. London Math. Soc. (2) **39** (1989), no. 3, 436–456, DOI 10.1112/jlms/s2-39.3.436. MR1002456

[116] Jan-Erik Roos, *Homology of loop spaces and of local rings*, 18th Scandinavian Congress of Mathematicians (Aarhus, 1980), Progr. Math., vol. 11, Birkhäuser, Boston, Mass., 1981, pp. 441–468. MR633376

[117] G. Scheja and U. Storch, *Lokale Verzweigungstheorie* (German), Schriftenreihe des Mathematischen Institutes der Universität Freiburg, No. 5, Institut des Mathématiques, Université de Fribourg, Fribourg, 1974. Vorlesungen über Kommutative Algebra (Wintersemester 1973/74). MR0412167

[118] S. Paul Smith, *Category equivalences involving graded modules over path algebras of quivers*, Adv. Math. **230** (2012), no. 4-6, 1780–1810, DOI 10.1016/j.aim.2012.03.031. MR2927354

[119] Øyvind Solberg, *A graded ring of finite Cohen-Macaulay type*, Comm. Algebra **16** (1988), no. 10, 2121–2124, DOI 10.1080/00927878808823681. MR956160

[120] Øyvind Solberg, *Hypersurface singularities of finite Cohen-Macaulay type*, Proc. London Math. Soc. (3) **58** (1989), no. 2, 258–280, DOI 10.1112/plms/s3-58.2.258. MR977477

[121] Gonçalo Tabuada, *$\mathbb{A}^1$-homotopy invariants of dg orbit categories*, J. Algebra **434** (2015), 169–192, DOI 10.1016/j.jalgebra.2015.03.028. MR3342391

[122] Ryo Takahashi, *On the category of modules of Gorenstein dimension zero*, Math. Z. **251** (2005), no. 2, 249–256, DOI 10.1007/s00209-005-0795-8. MR2191025

[123] Oana Veliche, *Gorenstein projective dimension for complexes*, Trans. Amer. Math. Soc. **358** (2006), no. 3, 1257–1283, DOI 10.1090/S0002-9947-05-03771-2. MR2187653

[124] Jean-Louis Verdier, *Des catégories dérivées des catégories abéliennes* (French, with French summary), Astérisque **239** (1996), xii+253 pp. (1997). With a preface by Luc Illusie; Edited and with a note by Georges Maltsiniotis. MR1453167

[125] Hsin-Ju Wang, *On the Fitting ideals in free resolutions*, Michigan Math. J. **41** (1994), no. 3, 587–608, DOI 10.1307/mmj/1029005082. MR1297711

[126] Hsin-Ju Wang, *On the Jacobian ideals of affine algebras*, Comm. Algebra **26** (1998), no. 5, 1577–1580, DOI 10.1080/00927879808826222. MR1622431

[127] Zhengfang Wang, *Singular equivalence of Morita type with level*, J. Algebra **439** (2015), 245–269, DOI 10.1016/j.jalgebra.2015.05.012. MR3373371

[128] Yuji Yoshino, *Cohen-Macaulay modules over Cohen-Macaulay rings*, London Mathematical Society Lecture Note Series, vol. 146, Cambridge University Press, Cambridge, 1990, DOI 10.1017/CBO9780511600685. MR1079937

Glossary

$X[1]$	Translation of an object X in a triangulated category. 7
$M(\delta)$	Shift of a graded module. 75
M^*	Right dual of a module M. 18, 76
${}^*M$	Left dual of a module M. 60
$M^\sharp$	Graded K-dual of M in $\mathrm{D}^-(\mathrm{Mod}(U(L)))$ as a complex of right $U(L)$-modules. 80
$-\underline{\otimes}^{\mathsf{L}}_S L$	Stable tensor product functor $\underline{\mathrm{D}}^{\mathrm{b}}(S) \to \underline{\mathrm{D}}^{\mathrm{b}}(T)$ for an S-T-bimodule L. 49
f^*	Exact functor $\mathrm{D}(T) \to \mathrm{D}(S)$ induced from $f\colon T \to S$. 50
$\underline{f}^*$	Exact functor $\underline{\mathrm{D}}^{\mathrm{b}}(T) \to \underline{\mathrm{D}}^{\mathrm{b}}(S)$ induced from $f\colon T \to S$. 50
f_*	Exact functor $\mathrm{D}(S) \to \mathrm{D}(T)$ induced from $f\colon T \to S$. 50
$\underline{f}_*$	Exact functor $\underline{\mathrm{D}}^{\mathrm{b}}(S) \to \underline{\mathrm{D}}^{\mathrm{b}}(T)$ induced from $f\colon T \to S$. 50
$f^!$	Exact functor $\mathrm{D}(T) \to \mathrm{D}(S)$ induced from $f\colon T \to S$. 50
$\underline{f}^!$	Exact functor $\underline{\mathrm{D}}^{\mathrm{b}}(T) \to \underline{\mathrm{D}}^{\mathrm{b}}(S)$ induced from $f\colon T \to S$. 50
B^{e}_A	Enveloping algebra of $A \to B$. 63
$B^{(\mathrm{e})}_A$	Derived enveloping algebra of $A \to B$. 158
$\langle\ ,\ \rangle_A$	Non-degenerate bilinear pairing $\mathrm{Ext}^r_R(A,R) \otimes_R A \to E$. 59
L^+	Sub-Lie-algebra of a super-Lie-algebra L spanned by homogeneous elements of even degree. 77
L^-	Submodule of a super-Lie-algebra L spanned by homogeneous elements of odd degree. 77
$\bigwedge V$	Exterior algebra of a free module V. 171
S^{op}	Opposite ring of a ring S. 18
${}_\varepsilon S^{\mathrm{op}}$	Graded ε-opposite ring of a graded ring S. 76
$\mathbf{a}$	Sheafification functor $\mathrm{mod}(U) \to \mathrm{Proj}\text{-}U$. 90
$\mathbf{A}(V,W)$	Affine super-space to the linear super-space (V,W). 90
$\mathrm{ann}\, M$	Annihilator of a module M. 171
$\mathrm{ann}_R\, \underline{\mathrm{D}}^{\mathrm{b}}(S)$	Annihilator of the stable derived category of S. 157
$\underline{\mathrm{APC}}(S)$	Homotopy category of acyclic complexes of finitely generated projective S-modules. 13
$\mathrm{art}(S)$	Subcategory of $\mathrm{mod}(S)$ of all artinian modules. 90
$\mathbf{b}_{(V,W)}$	Exact functor $\mathrm{D}^{\mathrm{b}}(\mathbf{U}(V,W)) \to \mathrm{D}^{\mathrm{b}}(\mathbf{U}(\mathbf{D}(V,W)))$. 99
$\mathbf{C}$	Co-resolution functor $\mathrm{D}^{\mathrm{b}}(S) \to \mathrm{K}^{+,\mathrm{b}}(S)$. 34
$c^i(X,Y)$	Induced morphism $\mathrm{Ext}^i_S(X,Y) \to \underline{\mathrm{Ext}}^i_S(X,Y)$. 39
$\mathrm{cl}(X)$	Class in $\mathsf{K}(\mathsf{C})$ of an object X. 171

$\mathrm{Cliff}_S(Q)$	Clifford-algebra of a quadratic form $Q\colon S^n \to S$. 141
$\mathrm{Cok}(f)$	Cokernel of a map f. 171
$\mathbf{CR}$	Complete resolution functor $\underline{\mathrm{MCM}}(S) \to \underline{\mathrm{APC}}(S)$. 21
	Complete resolution functor $\mathrm{D^b}(S) \to \underline{\mathrm{APC}}(S)$. 33
$\mathbf{D}$	Duality functor on the category of linear super-spaces. 88
$\mathbf{d}(N)$	Associated total complex of a bigraded module N. 100
$\mathbf{d}(X,Y)$	Morphism $\mathrm{cone}(\mathbf{N}(X,Y)) \to \mathrm{Hom}_S(\mathbf{CR}(X), \mathbf{CR}(Y))$. 39
d_X^i	Differential of the complex X mapping $X^{i-1} \to X^i$. 7
$\mathrm{D}(S)$	Derived category of $\mathrm{Mod}(S)$ of complexes with finitely generated cohomology. 7
$\mathrm{D_{gr}}(S)$	Derived category of $\mathrm{grMod}(S)$ of complexes with finitely generated cohomology. 75
$\mathrm{D_{sg}}(X)$	Singularity category of a scheme X. 145
$\mathrm{D}^-(S)$	Derived category of $\mathrm{Mod}(S)$ of complexes with finitely generated and bounded above cohomology. 7
$\mathrm{D^b}(S)$	Derived category of $\mathrm{Mod}(S)$ of complexes with finitely generated total cohomology. 7
$\mathrm{D^b_{a/p}}(U)$	Full subcategory of $\mathrm{D^b}(U)$ of all perfect complexes with artinian cohomology. 93
$\mathrm{D^b_{art}}(U)$	Full subcategory of $\mathrm{D^b}(U)$ of complexes with artinian cohomology. 92
$\mathrm{D^b_{gr_{\geqslant i}}}(A)$	Bounded derived category of graded A-modules concentrated in internal degrees $\geq i$. 145
$\mathrm{D^b_{perf}}(S)$	Full subcategory of $\mathrm{D^b}(S)$ of perfect complexes. 9
$\mathrm{D^b_{perf}}(\mathrm{Proj}\text{-}U)$	$\mathrm{D^b_{perf}}(U)/\,\mathrm{D^b_{a/p}}(U)$. 96
$\underline{\mathrm{D}}^{\mathrm{b}}(S)$	Stable derived category of S. 10
$\underline{\mathrm{D}}^{\mathrm{b}}_{\mathrm{gr}}(S)$	Stable derived category of graded S-modules. 75
$\det(V)$	Highest non-vanishing exterior power of V. 77
$\det(F)$	Super-determinant of a module or a morphism F. 79, 87
ev_A	Evaluation map $\mathrm{Hom}_R(A,E) \otimes_R A \to E$. 59
$\mathrm{Ext}_S(X,Y)$	Ext-module of S-complexes X and Y. 9
$\underline{\mathrm{Ext}}_S(M,N)$	Tate cohomology group of M with values in N over S. 37
$\underline{\mathrm{Ext}}_S^+(M,N)$	Even Tate cohomology group. 120
$\underline{\mathrm{Ext}}_S^-(M,N)$	Odd Tate cohomology group. 120
$\mathrm{Ext}_{\mathrm{GP}(S)}(M,N)$	Relative cohomology over S. 151
$\widehat{\mathrm{Ext}}_S(M,N)$	Tate cohomology of M of finite G-dimension with values in N. 151
	Stable Cohomology group of (M,N). 152
$\underline{\mathrm{fpd}}(S)$	Stable category of S-modules of finite projective dimension. 31
$\mathbf{G}(M)$	Resolution of M by G-projectives. 151
$\mathrm{G}_i(S)$	Algebraic K-group of finitely generated S-modules. 25
$\mathrm{G}\text{-}\mathrm{dim}_S(M)$	G-dimension of an S-module M. 151
$\mathrm{G}\text{-}\mathrm{mod}(S)$	Full subcategory of $\mathrm{mod}(S)$ of modules of finite G-dimension. 151
$\mathrm{GP}(S)$	Subcategory of $\mathrm{mod}(S)$ of G-projective modules. 133
$\underline{\mathrm{GP}}(S)$	Stable category of $\mathrm{GP}(S)$. 133
$\mathrm{gr}(S)$	Associated graded ring of S. 74

$\mathrm{gr}(M)$	Associated graded $\mathrm{gr}(S)$-module of the S-module M. 74
$\mathrm{grmod}(S)$	Category of finitely generated graded right S-modules. 7
$\mathrm{g}\underline{\mathrm{rmod}}(S)$	Stable category of finitely generated graded right S-modules. 75
$\mathrm{grMod}(S)$	Category of graded right S-modules. 7
$\mathrm{grP}(S)$	Subcategory of $\mathrm{grMod}(S)$ of finitely generated projective graded S-modules. 7
$\mathbf{H}$	Hull of finite projective dimension $\underline{\mathrm{mod}}(S) \to \underline{\mathrm{fpd}}(S)$. 31
$\mathrm{h}(M)$	Herbrand quotient for a representation of a cyclic group. 121
$\mathrm{h}(M,N)$	Herbrand difference of (M,N). 121
$\widehat{\mathrm{H}}_R(M)$	Local Tate cohomology of a ring R with values in M in $\mathrm{D}^-(\mathrm{Mod}(U(L)))$. 115
$\widehat{\mathrm{H}}(G,M)$	Tate cohomology of a group G with values in an integral representation M. 72
$\widehat{\mathrm{H}}(L/K,M)$	Tate cohomology of a super-K-Lie algebra L with values in M. 79
$\underline{\mathrm{Hom}}_S(M,N)$	Morphisms $M \to N$ in $\underline{\mathrm{mod}}(S)$. 11
$\mathbf{i}(X)$	Induced morphism $\mathbf{P}(\mathbf{P}(X)^*)^* \to \mathbf{CR}(X)[1]$. 33
ι_S	Exact functor $\underline{\mathrm{MCM}}(S) \to \underline{\mathrm{D}}^{\mathrm{b}}(S)$. 19
	Exact functor $\underline{\mathrm{GP}}(S) \longrightarrow \underline{\mathrm{D}}^{\mathrm{b}}(S)$. 135
i_X	Lowest non-zero degree of $\mathrm{H}(X)$. 42
i_X^*	Lowest non-zero degree of $\mathrm{H}(\mathrm{RHom}_S(X,S))$. 42
$I_{B/A}$	Augmentation ideal of B_A^{e}. 63
$\mathrm{Im}(f)$	Image of a map f. 171
$\mathrm{ind}^+(F)$	Even index of a morphism $F\colon (V_1,V_2) \to (W_1,W_2)$. 88
$\mathrm{ind}^-(F)$	Odd index of a morphism $F\colon (V_1,V_2) \to (W_1,W_2)$. 88
$\mathrm{injdim}_S(M)$	Injective dimension of the S-module M. 171
$J(\underline{f})$	Jacobian ideal of $\underline{f} = f_1,\dots,f_m$. 68
$J(S)$	Jacobian ideal of a commutative notherian ring S. 158
$\mathrm{K}(S)$	Homotopy category of $\mathrm{Mod}(S)$ of complexes with finitely generated cohomology. 7
$\mathrm{K}_{\mathrm{tac}}(\mathrm{proj}\, S)$	Category of totally acyclic complexes of projectives. 134
$\mathrm{K}^{-,\mathrm{b}}(\mathrm{P}(S))$	Homotopy category of all bounded above complexes of finitely generated projective S-modules with bounded cohomology. 8
$\mathsf{K}_0(\mathsf{C})$	Grothendieck group of a triangulated category C. 92
$\mathsf{K}_i(S)$	Algebraic K-group of finitely generated projective S-modules. 25
$\underline{\mathsf{K}}_0(S)$	Grothendieck group of $\underline{\mathrm{MCM}}(S)$. 25
$\underline{\mathsf{K}}_i(\mathbf{U}(V,W))$	Stable K-groups for $\mathbf{U}(V,W)$. 25
$\mathsf{K}'_i(S)$	See $\mathsf{G}_i(S)$. 25
$\mathbf{M}$	MCM-approximation functor $\underline{\mathrm{mod}}(S) \to \underline{\mathrm{MCM}}(S)$. 31, 34
m_X	Maximum non-zero degree of $\mathrm{H}(X)$. 34
m_X^*	Maximum non-zero degree of $\mathrm{H}(\mathrm{RHom}_S(X,S))$. 34
$\mathrm{maxspec}(T)$	Set of all maximal ideals of a commutative ring T. 59
$\mathrm{MCM}(S)$	Category of maximal Cohen–Macaulay S-modules. 18
$\underline{\mathrm{MCM}}(S)$	Stable category of maximal Cohen–Macaulay S-modules. 18
$\mathrm{mod}(S)$	Category of finitely generated right S-modules. 7

$\underline{\mathrm{mod}}(S)$	Stable category of finitely generated S-modules. 11
$\mathrm{mod}(K \times K)$	Category of linear super-spaces over the field K. 87
$\mathrm{Mod}(S)$	Category of right S-modules. 7
$\mathrm{Mod}_{lf}(U)$	Category of graded U-modules which are finite for the total degree. 89
$\underline{\mathrm{Mon}}(U)$	Homotopy category of all monads over U. 102
$\underline{\mathrm{Mon}}^0(U)$	Full subcategory of $\underline{\mathrm{Mon}}(U)$ of linear monads. 102
$\mathbf{N}(X)$	Norm map $\mathbf{P}(X) \to \mathbf{P}(\mathbf{P}(X)^*)^*$. 33
$\mathbf{N}(X,Y)$	Norm map $\mathrm{Hom}_S(\mathbf{C}(X), \mathbf{P}(Y)) \to \mathrm{Hom}_S(\mathbf{P}(X), \mathbf{C}(Y))$. 39
Ω_S	Loop-space functor $\underline{\mathrm{mod}}(S) \to \underline{\mathrm{mod}}(S)$. 11
Ω_0	Exact functor $\underline{\mathrm{APC}}(S) \to \underline{\mathrm{MCM}}(S)$. 19, 134
$\Omega_i(X)$	i-th syzygy modules of a complex X. 13
ω_S	Dualising module for a ring S. 19
$\omega_{S/T}$	Relative dualizing module for a ring map $T \to S$. 56, 58
$\mathcal{O}$	Structure sheaf. 171
$\mathbf{P}$	Projective resolution functor $\mathrm{D}^{\mathrm{b}}(S) \to \mathrm{K}^{-,\mathrm{b}}(\mathrm{P}(S))$. 20
p_M	Surjection from a projective cover $\mathrm{P}_M \to M$ for a module M. 11
P_M	Projective cover of a module M. 11
$\mathrm{P}(S)$	Subcategory of $\mathrm{Mod}(S)$ of finitely generated projective S-modules. 7
$\mathbb{P}(V)$	Projective space of hyperplanes in V. 87
$\mathbb{P}(V, W)$	Projective super-space attached to the linear super-space (V, W). 90
$\mathrm{Pic}(X)$	Picard group on X. 171
$\underline{\mathrm{Prim}}(U)$	Category of primitive objects over $\mathrm{Proj}\text{-}U$ for a graded ring U. 96
$\mathrm{Proj}\text{-}U$	$\mathrm{mod}(U)/\mathrm{art}(U)$ for a graded ring U. 90
$\mathrm{Proj}(S)$	Projective variety associated to the graded ring S. 171
$\mathbf{r}(X)$	Induced morphism $\mathbf{CR}(X) \to \mathbf{P}(X)$. 33
$\mathrm{rad}(S)$	Radical of the ring S. 71
$\underline{\mathrm{RHom}}(L, -)$	Stable Hom functor $\underline{\mathrm{D}}^{\mathrm{b}}(T) \to \underline{\mathrm{D}}^{\mathrm{b}}(S)$ for an S-T-bimodule L. 49
$\underline{\mathrm{RHom}}(-,-)$	Exact bifunctor $\mathrm{D}^-(\mathrm{Mod}(S)) \times \underline{\mathrm{D}}^{\mathrm{b}}(S)^{\mathrm{op}} \to \mathrm{D}(\mathsf{Ab})$. 51
$\sigma_{\leqslant k}X$	Naive truncation of a complex X. 14
$\underline{\sigma}_{\leqslant k}$	Exact functor $\underline{\mathrm{APC}}(S) \longrightarrow \underline{\mathrm{D}}^{\mathrm{b}}(S)$. 14
$\underline{\sigma}_{\leqslant}$	Exact functor $\underline{\mathrm{APC}}(S) \to \underline{\mathrm{D}}^{\mathrm{b}}(S)$. 15
$\mathrm{soc}(S)$	Socle of the ring S. 71
$\mathrm{Sym}(V)$	Symmetric algebra of a free module V. 171
T	Translation functor of a triangulated category. 7
τ_N	Trace map of the module N. 62, 117
$\vartheta^0_D(g)$	Dedekind different of $g\colon A \to B$. 66
$\vartheta^0_K(g)$	Kähler different of $g\colon A \to B$. 66
$\vartheta^0_N(g)$	Noether different of $g\colon A \to B$. 64
$\vartheta^0_{dN}(S/R)$	Derived Noether different of S over R. 158

U	Functor from the category of linear super-spaces to the category ofgraded Hopf-algebras with inversion. 88
$U(L)$	Universal enveloping algebra of the super-Lie-algebra L. 77
vdim(S)	Virtual dimension of S. 17
Z(B)	Centre of a ring B. 63

Index

SELECTED PUBLISHED TITLES IN THIS SERIES